HOMO
ANGELICANSIS

HOMO
ANGELICANSIS

Spiritual anthropology and the evolution of the soul

Izak J. Botha

South Africa

2007

First published 2007 by Izak J. Botha
P. O. Box 478, Sea Point, 8060, Cape Town, South Africa
http://homoangelicansis.com

Printed and bound by Creda Communications
12 Eliot Avenue, Epping 2, 7460, Cape Town, South Africa

A catalogue record for this book is available from the National Library, Pretoria, South Africa

ISBN-10: 0-620-37376-0
ISBN-13: 978-0-620-37376-0

This book is dedicated to Maria Magdalena Swanepoel

Acknowledgements

I would like to thank the following people who partook in my journey: Robyn Wilkinson for her years of dedication and support with the layout and contents of the book; Ilse Bigalke for her assistance and proofreading; Lesley Rochat for her encouragement and help with the synopsis; John Coetzee for the many inspirational explorations long before the book took its form; Peter Nortjé for the years of night-owling and making a significant contribution to the change of direction earlier in my life.

Also those who played significant roles in my spiritual journey: Tessa van der Merwe, Engela Goosen, Patrick Oberem, Gert Fourie (deceased), Bob Harman (deceased) and Mariétte Carstens.

And to those who still persevere: Annaretha Coetzee, Frans and Mandy Coetzee, Jaco and Moreen Coetzee, Willem Immelman, Justin Bottaro, Costa Champanis, Bennie Geldenhuis, Leigh Trimmer, Mandy Millard, Pieter Carstens, Marais Kikillus, Hamish Huston, Lucille Hubinger, Candice Kalil, Anel Mostert and Johan Malherbe.

Izak Botha
Cape Town, 2007

Contents

PART 3
Homo Angelicansis

PART 4
The Science of Spirituality

PART 5
New Frontiers

Introduction

Would you agree to having genetically engineered children? Moreover, if genetic engineering made them lesser beings than yourself, would you still agree? If you were to discover that you have evolved into a part mortal human and part immortal angelic being, would you want your children to remain mortal or would you want them to be at least as advanced as you are, empowered by immortality? If during the normal course of evolution we evolved to become spiritual beings with souls, would you want to be responsible for recreating new generations without souls?

Many humans live with the notion that we have a soul. The obtainment of a soul has always been contentious, the most common theory being that some or other "Creator" bestowed it on us. However, the evolutionary theory proclaims there has never been any interference by any such Creator. If that were so, we have to rely on the evolution of human biology to introduce a soul. Human biology would have to evolve in a spiritual direction on its own and, in the process, create a vehicle able to retain consciousness even after physical and biological death. The vehicle for consciousness to survive and continue its journey after biological death is called the soul.

Immortality through the soul will give humans the ability to cast off their physical nature and survive physiological death. The soul then, by implication, allows humans to continue their existence in a non-physical form, a manifestation known as spiritual. Spirituality is therefore a destiny, not a doctrine or belief. Being spiritual means you are closer to achieving conscious life without a physical body. Even though the non-physical destination of spirituality is yet ill-defined, humans have come a long way in establishing what and where it is. Although many attempts have been made in religious and esoteric practice as well as scientific discovery, nobody can yet say exactly how spirituality is achieved.

As far as we know, Homo sapiens is the single most evolved living entity in the entire universe. During the more than 14 billion years that our universe has been in existence, there is simply no evidence of a more complex or evolved creature than the human being. After an entire evolutionary process, from the big bang to the forming and separation of matter, to the creation of galaxies and the creation of solar systems, right up to the forming of life on Earth – the only life we know of – humans in the form of Homo sapiens are at the pinnacle of all creation. Well, until quite recently, that is! Let me show you how Homo sapiens is no longer the highest form of life; but has been surpassed by *Homo angelicansis*. We can describe Homo angelicansis as a half-breed, the hybrid product of two distinct types of beings and a cross between physical and spiritual natures.

Although the event marking the point when humans acquired souls has long been postulated, the evolutionary significance of this extraordinary incident has almost gone unnoticed. The current theory of evolution does not make provision for spiritual life and the possible existence of a human soul. Nevertheless, the concept of evolution is still so new and unexplored that we may yet find a credible place for spirituality and soul in its synopsis. Sometime during evolution, life became conscious and that consciousness became multi-dimensional. Humans are no longer only physical creatures, but have also become spiritual beings with a spirituality that can now be defined.

How ironic that just as we are now at that point in history where we are able to define our spiritual nature and the soul, we can also destroy precisely that which we have taken hundreds of thousands, if not millions, of years to become. We can destroy the soul. Standing at the threshold of our bio-technological age, with the ability to "play God", we are able to recreate or mimic physiological life. But, we have not yet discovered how to create any life with soul. In our quest to clone and replace body parts with artificial parts and artificial intelligence or even create cyborgs, we have forgotten one aspect. The soul! How do we imitate and recreate a soul? This document is my theory on the soul: What it is, where it came from and its destination, or shall we say its former destination.

> *Spirituality is the consciousness humans display without a physical body. We assume the vehicle to be the soul.*

Almost all humans at some time during their lives search for spirituality. Whether this search is of a godly or demonic nature or conducted in a religious or esoteric way does not matter. Any search seeks to deepen understanding and establish the significance of a spiritual life. Spirituality validates the human experience and provides meaning to everyday lives. After all, how do we justify a limited lifespan on Earth without knowing we have a soul to carry our experience further after death?

Whether spirituality plays a greater or lesser part as we evolve as human beings is debatable and deserves some investigation. For many, it has become a mere religious experience, shaped and determined by unyielding doctrine. However, spirituality is more than just "being religious". In previous centuries, attempts made by various philosophers and theologians at defining spirituality were based upon observations made by individuals at an early stage of human consciousness. In an examination of these age-old perceptions, we see how we can add to them in the light of new discoveries, higher intelligence and greater levels of consciousness.

Some of the most prominent aspects of human nature distinguishing us from other animal species are that we have sophisticated technology, conscious awareness, the ability to communicate using language, and we seek to be spiritual. Some would include higher intelligence in this comparison but as intelligence has yet to find an accurate definition, I prefer to leave it out. Notably, we are uniquely identifiable in our ability to express ourselves through the notion of an objective soul.

In coming to terms with mortality and the finality of death, we are destined to question the meaning of life. Existing as conscious and thinking animals for a short period before dying, barely to be remembered thereafter, provides little justification for living. So why do we? Perhaps our constant struggle for survival comes from instinct alone. Although survival on Earth is of a physical nature, the conscious part of us desires to extend life after physical or biological death. At present, we rely on the notion of a soul to entertain the idea of continued life.

When spirituality is incorporated into religious teachings and religious doctrine, it becomes mere conviction. Religions are static and unchanging and rely on *res adjudicata* status for their survival. The Bible according to Christianity is supposed to be the sole word of God and cannot ever be altered. That means that for the remainder of human kind's existence, be it millions if not billions of years, or perhaps eternally, we must abide by unchanging scriptures based on the insights of people who lived thousands of years earlier when speech and writing first came into existence. Thus, we are governed by the doctrines, visions and perceptions of people who lived thousands of years ago and were only really starting to evolve as conscious beings.

This primitive viewpoint may be an indication of how little we have evolved as a species. We have become incredibly intelligent, but spiritually deficient. How can we claim to be highly evolved if we are spiritually still thinking the same way humans did thousands of years ago? Quite simply, we do not. We are everything but evolved in the greater scheme of things. Because we have become highly intelligent compared to archaic civilizations by increasing our capacity to think logically and improving our intelligence and emotional quotient, this does not mean we are more evolved, or have deeper insight into matters related to spiritual life.

Suzan Blackmore in her book *The Meme Machine* (p 203) expresses this point in a nutshell:

"Religions build theories about the world and then prevent them from being tested. Religions provide nice, appealing and comforting ideas, and cloak them in a mask of 'truth, beauty, and goodness'. The theories can then thrive in spite of being untrue, ugly, or cruel.

"In the end, there is no ultimate truth to be found and locked up forever, but there are more or less truthful theories and better or worse predictions. I do defend the idea that science, at its best, is more truthful than religion."

Our deficiency regarding spirituality and soul-related matters is obvious. As thinking animals, we have immortalised matters relating to spirituality and the soul's existence. This has caused a huge diversion in our progress on the subject of spirituality. There is no place for spiritual evolution or soul survival of the fittest in religion. Spirituality has been watered down to mere superstition and become doctrinal, naïvely followed by billions of unsuspecting and unenlightened souls.

We do not know for certain whether there is a godly dimension or realm in which a "God" presides and creates our reality. We also do not know whether the soul comes from or is on its way to such a godly dimension. We cannot scientifically prove such reality exists. What we can say is that for some strange reason we feel a connection with spirituality and some entity we call the soul. Certainly, spirituality and the soul mean different things to different people and societies. In some circles, such as the humanists, naturalists and scientific materialists (those who exclude supernatural or spiritual phenomena), there is much doubt whether the soul exists at all. To say there is a soul is therefore an individual sentiment, even if we find people with similar views who choose to believe it exists in some form or another. None of this is proven scientifically, which means we are left with either a belief system or an individual perception.

To make the life of our soul less supernatural and spirituality less religious, one has to analyse its makeup and existence more carefully and examine whether there is any substantial proof of its existence, not forgetting that proof may not necessarily mean scientific proof, but may also include such. After all, the theory of evolution and that of the big bang are by no means proven yet, but we direct our lives and scientific research as if they are. Continuing evidence for such theories has become increasingly overwhelming, and until proven otherwise, they remain credible.

One reason we focus on a spiritual or soul nature is human survival. As a species we have the desire to survive after earthly life – we even speak about pioneering alternative planets or solar systems. We also express the need to continue conscious awareness in the afterlife, a way by which we can survive physical death via the soul. Humans were not conscious of evolution for the greater part of their own evolution. Only in the last few centuries have we become aware of biological change occurring over generations of species. Charles Darwin (1809-82) in *The Origin of Species* made the evolutionary theory famous in the 19th century. For billions of years the universe unfolded, galaxies formed, life came into existence and evolved to its current complex state. All this happened without any assistance from our consciousness. There simply was no known consciousness from any living species with any awareness of the process of evolution before we developed our own over the last 200 000 years. It is only now, some 14 billion light-years after the universe appears to have originated, that we are able to discover and study our past.

As evolved as we have become, we are still not conscious of further and more complex evolutionary processes taking place such as prospective spiritual evolution. In my opinion we are still not developed enough and conscious enough to make such observations. How much do we know about our evolution into soul beings? If the soul exists, it seems that the true purpose is still hidden. We need to find a new perspective or possible grounds for its existence. Instead of assuming lofty positions regarding how well we know "God" and "God's will", we need to consider our infancy relating to spiritual matters and a possible soul existence.

The misperception that spirituality has always been present stems from the creationist's doctrine of God first creating the Earth, then man, who immediately thereafter receives a soul, as if everything came into being in one act. Another view stemming from Hinduism maintains that the soul existed first and is now acquiring material form or a physical nature. Occultism teaches how biological life-forms evolved separately from something akin to a "soul society" and that humans were invaded or entered by the souls once human nature had evolved sufficiently. (I deal with these scenarios in more depth in the first half of the book). According to esoteric views, soul beings first existed alongside human forms and then eventually entered the developed human entities when they were ready for such invasion. This is supposedly achieved during the embryo phase in the mother's womb, but it does not explain the origin of these souls. This theory sounds like a polite way of portraying the crude manner in which humans are possessed by soul creatures from alien realms, forcefully occupying bodies like viruses or parasites.

The above perceptions are negated by the scientific view derived from discoveries in cosmology and anthropology based on the big bang and the evolutionary theories. Here we have a wealth of evidence that our universe has been in existence for more than 14 billion years, life on Earth has been around for more than 3,5 billion years and our own human nature evolved from primates to Homo sapiens over approximately the last 5-6 million years. For the evolutionists, life on Earth started purely by chance and managed to evolve into the incredibly complex physical anatomy with the intelligence and consciousness we currently have. However, according to some of our leading scientists, humans may not actually have a soul.

To my mind, there is a measure of truth in all these theories, but once again, they all fall short of the mark. The soul may not have existed for as long as Homo sapiens has been on Earth as much as it may not have been bestowed onto biblical Adam 6 000 years ago. When we consider evolution, we see that the so-called "spiritual nature" of the universe and a "soul nature" of life forms on Earth may not have existed from the beginning of creation. The possible existence of both may have been later additions to our physical nature. What is more, the evolution of our soul could be the final ontogeny of human nature. Rather than the soul invading or giving life to the human race, we may be giving life to the soul in order to have a vehicle to travel to other realms. The soul may be evolving into a formidable entity that can survive a holocaustic future or other calamity that humans and all other life forms may encounter. The soul's purpose may even be for intergalactic migration. Perhaps life forms will migrate to Mars or even other galaxies in a non-physical body such as the prospective soul. If not Mars or further into space, why not another dimension altogether with this spiritual vehicle that does not require a space suit?

What came first? Did the spiritual realm and soul exist first, or did the big bang genesis, our physical universe and life forms come about before such celestial existence? This question leads us to ask what spirituality really is. Spirituality is immensely difficult, if not impossible, to define. Historical records of burials, paintings and written documents show that we have been recognising our spiritual nature for tens of thousands of years and yet we have no conclusive answer.

Perhaps the best place to start is to deal with what spirituality is not. In my opinion, spirituality does not entail being religious or following a religious creed as practised by Christianity, Buddhism, Hinduism, Islam or any of the other religions. I do not speak of spirituality in the context of a relationship with disembodied beings, godly angels (good spirits) or demons (evil spirits). For me, spiritual expression does not mean receiving messages or communicating with spiritual entities or beings. It does not entail interaction with a spiritual hierarchy that refers to the series of spiritual beings envisioned as the various orders from God to man. Lastly, spirituality to me has nothing to do with modern views on "spiritualism", seen by some as a belief that the dead manifest themselves to humans through a clairvoyant, medium or other source. Spiritualism in its commonly accepted meaning refers to teachings and practices received by communicating with spirits of the dead through mediums or séances. This is not what I am referring to or how I experience spirituality and I will therefore not be drawing my inspiration from such sources.

One would think there would be consensus regarding the inception of our soul nature, but that is not the case at all. Amongst those who presume there is a soul, there are three main opinions. Some say the soul existed before physical or biological life; some that it is the result of extraterrestrial interference; and some say that humans existed first and the soul was a later addition. There are obviously also those who think that the soul does not exist at all, but since my work revolves around the hypothesis of the soul's existence, I will not pursue this argument in detail. By looking further into the minutiae of these opposing views, rather than debating whether the soul existed before biological life or that the biological being was first created and implanted with a soul, we can take this study one step further and examine overall perceptions of what the soul may or may not be.

According to Eastern beliefs and some Western groups, souls wanted physical experience and so chose to incarnate on Earth as humans or through humans. In certain religions, this is a once-off happening and for others it occurs more often as explained by the theory of reincarnation. Some religions such as Hinduism believe we can reincarnate in forms other than human. Modern views like those of Michael Newton (counselling psychologist, Ph.D., author of *Journey of Souls*, 1998 and *Destiny of Souls*, 2001) cite case studies with hypnotised patients. He suggests there is a hierarchy of souls in an alternative realm using earthly human bodies to acquire physical experience. This theory is no different from occult views of the past hundreds if not thousands of years. The essence of their teachings is that the soul nature requires physical or material life experience and so takes on matter as a new form, or invades already existing life forms like humans to gain a body for physical experience.

If we leave organised religion aside for a moment and look at science, we find studies solely concentrated around physio-biological life and intelligence, but these rarely include the concept of soul. There are two schools of thought in this field. On the one hand, we have scientists going out of their way to prove that all human experience and our associated consciousness is the result of neurological actions in the brain that give rise to the mind, and nothing more. Our science determines all life resulted from the spontaneous assembly of molecules that form cells, followed by a spontaneous assembly of the genetic code. This code determines the form of the individual entity and initiates evolution towards the present-day consciousness produced in the human brain. The other view is esoteric, which we equate with occult science. Esotericism searches for an alternative source to which we can attribute cognitive experiences in humans by hypothesising that not all consciousness is merely the product of neurological firing from electrochemical reactions in the brain, but that there is some informative force making consciousness possible. This is a vast step closer to the possibility that soon we may be able to include spiritual experience in our quest for understanding the complete nature of life.

Everything about spirituality revolves around the soul. Spirituality and soul life are synonymous. Are soul and spirit the same? Is there a scientific purpose for the soul? In the following chapters, I set out the different rationales for the soul and outline different theories and hypotheses. I also make the distinction between soul and spirit but do not continue explaining the spirit other than to name its place according to previously established views – this book is not about the spirit or what it may or may not be. Although I differentiate between spirit and soul, we will concentrate on the soul and its existence. The cardinal importance of establishing when and how we may, or may not, have acquired a soul will be outlined. In my opinion, it came only after physiological life forms were well established and evolved. In the second half of the book I give reasons for the development of the soul representing the next step in our evolutionary advancement.

On the question of whether there is a spiritual realm, which is closely connected to the soul realm, I relied heavily on scientific discoveries of the last 150 years or so. In doing so, I examined breakthroughs in physical sciences and the ideas presented by neurosciences, to establish a link between our conscious experience of spiritual reality and dimension. Finally, I show how our non-religious perception of the soul is well established and that it is currently flourishing as at no other time in the recorded history of life on Earth. The hypothesis I present demonstrates how we can successfully incorporate the development of the soul in the natural process of evolution. The entire process being no longer buried in the unconscious, or even in the subconscious.

However, where does that leave spirituality? I will show that spirituality is the process by which humans acquire a spiritual vehicle, in the form of the soul, to retain conscious awareness even after corporal death. The soul exists in a non-material form in a realm we currently call the spiritual. Spirituality can also be called "spiritual anthropology", which revolves around the evolution of the soul and the preservation thereof.

How did I reach this conclusion? Read on ……

Homo angelicansis

PART 1

Myth, Ritual and Magic

Chapter 1

Atlantis and Lemuria

According to our documented and officially accepted history, urban civilization originated about 7-8 000 years ago with the Sumerians in Mesopotamia. The preceding 30 000 years leading up to this period have always been steeped in myth and superstition. New oceanic archaeological discoveries are shedding more light on past mysteries and it is bound to change the beginning of recorded history. Most of the early civilization myths refer to a period of great planetary upheaval. Scientists have discovered that major parts of the Earth were destroyed by cataclysmic events over a short period. These disasters are attributed to calamities such as meteorites, earthquakes, volcanic eruptions and flooding.

During the last ice age, a people called the Cro-Magnon represented Homo sapiens living in western and southern Europe. Skeletal remains of the Cro-Magnon were amongst others found in a rock shelter in the Dordogne of south-western France. The Cro-Magnons were advanced enough to have fitted clothes, jewellery and ornaments made from shell and bone. Their use of tools made from stone, bone and ivory and their colourful cave paintings are indicative of a complex culture. These discoveries point to an advanced cave dwelling civilization that existed before the Mesopotamian era.

Some of the early Sumerian texts and virtually all of the early mythology refer to lost civilizations of highly advanced almost supernatural people that came before Mesopotamia. Even though our archaeological discoveries of the Cro-Magnons have proven them to be an advanced people, we can say with confidence that they were not advanced enough to represent the supernatural culture referred to in the Sumerian myths. Either these almost apparitional civilizations of the Sumerian myths existed some time after the Cro-Magnons, or perhaps the Sumerians were really supernatural and lived alongside the rest of our human ancestors.

We now consider these mythical civilizations of Sumerian accounts to have existed more than 10 500 years ago. These lost ancient civilizations were destroyed by massive floods. Much scientific evidence shows that giant cataclysms changed the Earth between 17 and 8 000 years ago. Before this time, fully evolved human beings are thought to have existed, although most of the areas where they lived are now deeply embedded in the ocean. We have no certainty about what caused the floods and exactly when they took place, but there are two theories: The first is that the flooding was the result of melting ice towards the end of the last ice age. The other is that Earth was the victim of a meteorite that caused gigantic waves or tsunamis destroying the low-lying areas of land leaving only pockets of inhabitants. Both ice melting at the poles and a meteorite could cause sea levels to rise by more than 100 meters, creating floods of cataclysmic proportions.

At the beginning of the Pleistocene period or ice age (about 1,8 million years ago) ice sheets started building up in the Northern hemisphere causing glacial periods interspersed with warmer interglacial periods. The last of the ice existed until about 10 000 years ago with the start of the period we refer to as Holocene. According to Graham Hancock, author of the book *Underworld*, major destruction took place as a result of three global super floods more or less 15-14 000 years ago, 12-11 000 years ago and 8-7 000 years ago. Stories of floods are found in Vedic India, the America's, Mesopotamia, ancient Egypt, South East Asia, Northern Europe and across the Pacific. All of them mention that many people perished. Some researchers believe the myths may depict pure fantasy. Although many variations exist, historians do conclude that major floods could have led to the destruction of a prehistoric culture.

One of the stories about flooding is from the Babylonian Epic of Gilgamesh, which is an ancient text written on 12 clay tablets around 2 000 BC. Gilgamesh was a tyrannical Babylonian king who ruled the city of Erech (present day Iraq). This account is similar to the account of the flood in the Old Testament of the Bible. Genesis 7:17-24 speaks of floods lasting 40 days and the water prevailing for 150 days. According to this passage, the only survivors were Noah and his family. Erich Von Daniken, author of *The Return of the Gods*, points out that the flood story in the Bible is recorded in the second person, while in the Gilgamesh epic the first person is used throughout, suggesting an eyewitness account of someone who actually experienced the flood.

In Sumerian tablets on lost civilizations not only are the floods mentioned, but much is also said about lost civilizations that existed prior to these cataclysmic catastrophes. The Sumerians, who lived in the Persian Gulf, are considered the founders of the oldest urban civilization in the world. Their extraordinary and mysterious culture burgeoned during the fourth and third millennia BC. One of their surviving written versions of a flood was found when the city of Nippur on the Euphrates was excavated. This account seems to run parallel to that in the Old Testament involving Noah. Also found at Nippur were clay tablets dating as far back as the third and fourth century BC. These tablets refer to earlier civilizations and specifically to five antediluvian cities, which were apparently destroyed by floods. Although 37 lines at the beginning of the tablets are missing and we have no idea what it said, the bit that was recovered describes the creation of human beings, animals and plants. Following another missing piece, it describes a highly civilized era. During this time before the flood it says that "kingship was lowered from heaven", echoing similar stories of interaction with something from above seen in ancient Egyptian scriptures such as the *Pyramid Texts*, the *Book of What Is* in the *Duat* and the *Hermetica*. It appears that the inhabitants have evoked the gods' displeasure and that this results in an earth-destroying flood. Other parts of the tablets seem to divulge dissatisfaction from the gods regarding the decision to destroy the civilization. Then a man called Zisudra, a god-fearing king (similar to the biblical Noah), appears with seemingly the task of building some sort of vessel to survive the floods. Graham Hancock's *Underworld* gives an account of An and Enlil, two gods who were so grateful to Zisudra for building his Ark and surviving the flood that they decide to make him immortal.

It is at this point that I am forced to deal with mythology that speaks of supernatural gods. The same scriptures that tell us about super floods (of which we have scientific evidence), also mention the supernatural predecessors. Although we know the floods are real, how do we know the "gods" are also real?

So many discrepancies and uncertainties exist in the interpretation and in these mythological accounts that many historians afford them little credibility. What is not disputed, is that the tablets most certainly speak of an urban civilization somewhere in the Persian Gulf area with cities named Badtibira, Eridu, Larak, Sippar and Shurrupak, all of which were destroyed by a flood. The history of this region seems to be divided into periods before and after the flood.

Anyone growing up with old-fashioned Catholic or Christian values will remember lessons from the Bible containing the same principles of creation by a God and about Noah and the flood. The similarity is uncanny. However, the Bible stories are far more famous and accepted than any other because of the enormity of the western religious empires that revere them as scripture.

It is with this abstraction in mind that we start questioning the creation of the soul and its nature. Those brought up with western religion are indoctrinated into believing that the biblical account of creation was the start of man and the soul as an immaterial and life-giving force. Adam and Eve, followed by quite a number of generations, existed before Noah. In biblical context, Noah's ancestors from Adam onwards are regarded as early civilizations that already possessed souls. This would mean people with souls existed before the advent of the floods. However, as we see later, the first real evidence of the soul as a factor in the lives of early civilizations is recorded in documents from Egypt.

For ancient Egyptians, the "ka" or soul was recognized as the breath leaving the body when a person died. The Bible was written much later than the Egyptian texts (thousands of years later), leaving quite some discrepancy when the Bible speaks of the first man. Therefore we are faced with factual floods, mythical gods and contradictory interpretations of when, where and how the human race received souls. This topic in itself constitutes a vast subject and I will not explain it all here. Suffice to say it does become clearer later on when I discuss the history and evolution of our perceived soul.

Another discrepancy in the Biblical account of Adam as the first human to receive a soul when God breathed life into him (Genesis 2:7), is that we know other humans existed before and/or alongside the first Adam. Adam and Eve were not the first human beings, and whether they were the first to have souls, we do not know. Somehow there must have been a significant change in the way humans perceived themselves to fabricate a story that contains one of the early accounts of the soul becoming part of our nature.

David Rohl, author of *The Lost Testament*, maintains that although the chronology of the Old Testament in the Bible is wrong, the contents are true. The same goes for the floods that took place before Adam ended up in Eden. Rohl believes the floods took place before Adam and Eve and not afterwards as is written in the Bible. Archaeological evidence confirms the floods to have taken place more than 10 000 years ago. It would therefore make much more sense that the floods preceded Adam.

Interpreting it this way also enables us to match the other earlier sources on the floods with that of the Bible. Seen in reverse, the "first time awareness" of the soul from the Biblical viewpoint with Adam becoming the first person with a soul, would coincide with the "first time awareness" of the "ka" of the Egyptians. Both would date back about 5-6 000 years, which is thousands of years later than the super floods and well into the first urbanization of humankind.

Rohl believes that even if we accept the general dating in the Bible, we are unable to place Adam much before 5 400 BC. With science proving that civilized human life can be traced long before 6 000 or even 8 000 years ago, we clearly see that Adam was not the first physical human on Earth. In Genesis both Cain and Seth, Adam and Eve's offspring, took wives. These wives were not daughters of Adam and Eve, so where did they come from? Furthermore, Genesis 4:14 says that when Cain murdered Abel he feared he would be a fugitive sought to be slain by other humans if he was exiled.

We also now know that some ancient but highly developed civilizations existed thousands of years earlier than even the Sumerians, but were destroyed in a period of massive cataclysms. Two of these lost civilizations are said to be that of Atlantis and Lemuria. There is a genuine problem at this point. On the one hand, we have the still official version of history according to which our current civilization has its roots in Mesopotamia about 7-8 000 years ago, with full urbanization only dating back some 5-6 000 years ago and no higher evolved civilized people before this time. On the other, we have archaeological proof of highly developed civilizations that existed way before this time, but were destroyed by cataclysms some 10-12 000 years ago after the previous ice age. Any reference to the so-called "land of the gods" and "lost civilizations" were initially thought to be myth and not taken seriously by scientists and archaeologists who needed hard facts. Those hard facts are now starting to emerge. No official history has been recorded for the time between the extinction of the Neanderthals some 30-40 000 years ago with the invasion and dominance of the Cro-Magnons in Eurasia, and the beginning of modern civilization. The official history of urbanization starts with Mesopotamia, then Babylon and Egypt approximately 7 000 years ago, but the period before this is shrouded in controversy and uncertainty. We are only now starting to re-write history to include the lost civilizations that existed before our current official Mesopotamian history.

The lost civilizations known as Atlantis and Lemuria/Mu were considered mythological for the biggest part of our history. Now, with new evidence to the contrary, we are starting to think twice about the reality of these lost civilizations. Historians have shown that apart from Mesopotamia, other greatly advanced civilizations also appeared in Egypt and the east in what is now known as India.

David Icke, like many others, has another opinion. Icke, author of *Children of the Matrix*, believes Sumer was not the start of "civilized" society on Earth, but rather that that civilization originated in Atlantis and Lemuria. He is of the opinion that the Sumerian, Babylonian, Egyptian and the Indus Valley civilizations originated tens of thousands of years before our official history currently dates them. Icke also thinks the advanced Egyptian cultures, which "suddenly" appeared, were not independent of Sumer as claimed by historians, but part of Sumer and ruled by the same leader. Many elements of modern society, such as administration, law and building techniques, can be traced to Sumer. Icke refers to it as the Golden Age, a global society ruled by gods and their representatives who were highly advanced technologically. Arts, crafts and awe-inspiring architecture existed as evidence of a very well organized community that appeared over a short period. Whether an unexplained mystery like the pyramids was built before the cataclysms or by the Sumer Empire after the upheavals, is still being debated.

Atlantis

Plato in the *Timaeus* and *Critias* wrote about a lost civilization of Atlantis, but said nothing about Lemuria. Around 360 BC, Plato described an ancient island kingdom buried beneath the sea as a consequence of its inhabitants angering the gods through their corrupt and evil ways. Apparently an Egyptian priest described the island as being bigger than Asia Minor and Libya combined. This priest spoke about a burgeoning civilization that had conquered all the Mediterranean people except the Athenians. Plato outlined the nation as an utopian commonwealth and described buildings such as an acropolis with a vast palace and several temples surrounded by rings of land and water with bridges, canals, docks and a huge surrounding wall with adjacent housing. The city was surrounded by mountains with the island split into ten districts, each ruled separately. Metals used to cover walls included gold, silver, bronze and tin, and a metal referred to as "orichalc". Plato described waterways, docks with ships, horse racing tracks, public baths, gardens and temples filled with gold statues. Although his material may be fictional, it is also possible that he had access to accounts that no longer exist.

Plato, along with historians Allan and Delair in *When the Earth Nearly Died*, date the destruction of Atlantis to around 9 000 BC. After Plato's description, not much is said about Atlantis for more than 2 000 years until Ignatius Donnelly resurrected Atlantis during the 19th century. He believes Atlantis was the birthplace of modern humans where they developed before founding other early civilizations in Egypt and South America. Geographically he places Atlantis in the Atlantic Ocean. Other historians such as Ian Lawton in *Genesis Unveiled* comment that geological analysis may yet prove this realistic, but think Donnelly relies too heavily on coincidental and generalized similarity between the existence of Egyptian and South American pyramids and hieroglyphs. We now look at other supporters of the theory of Atlantis, although none of them are entirely credible.

French geologist and prehistorian Jacques Collina-Girard suggested that Atlantis was a small mid-channel island sitting in what is now the Strait of Gibraltar – swamped by rising seas along with six other islands. He places the islands anywhere from 53 to 125 meters below the ocean along the coast of Spain and Morocco. We do know that 20 000 years ago sea level was more than 122 meters lower than today.

Collina-Girard later suggested that Atlantis might be found where Plato originally suggested – on an island situated in front of the straits.

Another theorist is Jim Allen, who believes Atlantis is the Altiplano in South America, which consists of a series of basins between mountains between the eastern and western cordilleras of the Andes. A more popular view is that Atlantis was the Aegean island of Thira (north of Crete) that was destroyed by volcanic eruptions in 1470 BC. This view does not take into account Plato’s theory. Scholars have also identified the island with Crete, the Canary Islands, the Scandinavian Peninsula and America. In fact, initially the favoured theory was that the Americas themselves were Plato's Atlantis, but this was shifted to the Atlantic Ocean when found that native American ancestors more likely originated from the East. The latest 21st century contender for the lost city of Atlantis is the sea around Cyprus, where American researcher Robert Sarmast has recently discovered evidence of submerged man-made structures.

There is no sure winner yet!

Whether using geographical research, philosophical theories, archaeological findings or historical analysis we get different answers in every case. Assuming Plato based his writings on actual fact as opposed to literary license, there is still not enough to draw absolute conclusions. The Greek island of Thera does not match Plato’s location or timescale, neither does Atlantis in the North Sea as Atland or Aldland, and theories surrounding the continent of Antartica are geologically impossible. One of the more plausible ideas is a submerged landmass in the Caribbean Sea or that a large continent in the Atlantic Ocean was submerged. Then again, that does not fit the currently dominant theory of continental drift.

Despite large discrepancies regarding the geographical location of Atlantis, we also need to look at whether there is any recorded consistency regarding the nature of the people and details of their destruction. A lot of material exists that has been “telepathically” transmitted.

I think it is worth looking at such material while keeping in mind that there is ample scope for delusion.

Lemuria

Apart from Atlantis, there is also Mu, or modern day Lemuria. There is some controversy as to exactly how the name originated. Spanish priest Diego de Landa, after burning many of the original Mayan books, became an avid scholar of Mayan tradition. He believed the Mayan ancestors were the ten lost tribes of Israel. He also managed to obtain a false translation of the alphabet by frightening the natives. When Charles-Etienne Brasseur de Bourbourg discovered the alphabet, he used it to translate the Troano Codex and produced a grand tale of a lost continent that is considered to be largely fabricated. He also stretched his translation by attributing the name Mu to two symbols he was unable to translate.

The legend of Mu is found on islands all over the Pacific Ocean. Polynesians tell a story about a continent in the Pacific that was the motherland of all humankind. Another theory about where the name Mu came from, is connected to the zoologist Philip L Sclater who used it to explain the distribution of lemurs, a small primate found in Africa, Madagascar, India and the East Indian archipelago. From a zoological perspective, Sclater asserted that a continent stretching from Madagascar to Ceylon and Sumatra must have existed prehistorically. Earnst Heinrich Haeckel, a German naturalist from the turn of the 19th century and champion of Darwin, used Lemuria to explain the absence of fossil remains of early man. He believes that if man originated on a sunken continent in the Indian Ocean, all the fossils of the missing link are now under the sea. Many writers have increasingly placed Lemuria in the Pacific Ocean, but geologists believe if this was so it must have sunk before humans existed. Today just about everyone who has heard of Lemuria assumes that the legends of Mu are identical to the English zoologist's land of the lemurs.

It is only recently, at the very end of the 20th century, that archaeologists started discovering lost civilizations in the form of temples and structures rendering the myths not so mythical after all. Recent archaeological finds in the Indian Ocean seem to fit the profile for Lemuria, since it is here that most underwater archaeological structures resembling temples are found.

There are three important oceanic archaeological finds of late that can be used as examples of what is now considered ancient Lemuria. Mahabalipuram in Tamil Nadu, South India, the U-shaped structure at Tranquebar-Poompuhur near India and finds near Japan's Ryukyu Islands.

Author Graham Hancock spent several years cataloguing and studying myths about a once large and beautiful city with seven temples submerged beneath the sea. The story goes that the city was so beautiful that the gods became jealous and sent a flood to cover it. The Dorset-based Scientific Exploration Society (SES) and marine archaeologists from India's National Institute of Oceanography followed up some of Hancock's theories and found these temples beneath the sea offshore from Mahabalipuram in Tamil Nadu, South India.

Just off the Tranquebar-Poompuhur coast at 23 meters below sea level a U-shaped architectural structure was discovered, perhaps the remains of an unknown culture. An oceanographic study from the Gulf of Kutch in north-western India shows that 10 000 years ago the sea level was 60 meters lower than it is today. If we use this calculation to date this structure, it might just be 11 000 years old. That means it is 6 000 years older than the first monumental architecture of ancient Egypt or of any significant architecture from ancient Sumer in Mesopotamia.

South West of Japan's Ryukyu Islands at a depth of 30 meters on the ocean floor is a structure believed to be about 12 000 years old. Predating the pyramids of Giza by about 5 000 years, the terraced remains is some sort of man-made structure and a gigantic coral-encrusted pyramid-like monument. With a ceremonial centre with broad promenades and flanking pylons, the building measures about 110m long, 50m wide and 20m high. Our official history of human development does not offer any credible explanation of any living people from this time that could have built such a monument. Early Japanese civilization has been traced to the hunters and food-gatherers of the Neolithic period around 9 000 BC. It is not difficult to see why the early myths speak of the supernatural superseding civilization.

A study of the early civilizations helps us to determine whether their recorded mythology provides some explanation for the possibility of a soul nature. Alternatively, perhaps all this mythology is fantasy and born from the imagination of what, in today's terms, we consider relatively primitive and uneducated civilizations. On the other hand, can we deduce that spirituality is separate from mythology and the spiritual experience and possible existence of soul is actually factual? Unfortunately, everything about the soul in ancient text is questionable since it is saturated with myth, magic and the supernatural. The belief in the soul could be a figment of our imagination. Early civilizations used mythology to give meaning to the unexplained. With myth and mysticism as a backdrop, we have to try and understand the eschatology and nature of our soul. Do we really have souls or were we perhaps fooled by ancient myths and doctrines and therefore taken for a ride all along?

It is standard practice in all cultures to make use of myth to explain important philosophical concepts like the origin of man. There is a fine line between imagery and myth and we have no way of telling whether the myths have recorded actual happenings or whether they were purely fantasized. After all, myths are complex and refer to extraordinary times and places with supernatural beings and/or events. Often mirroring deep aspects of culture and human nature and important religious stories of creation, they document early traditions. Explaining different parts of the cycle of life, some myths deal with creation, others show how social order came about, and others deal with the end of the world through eschatology.

Modern-day Fantasies (Theories)

It is not my aim to prove or disprove the existence of these lost civilizations. Such debate would be lengthy and is not the purpose of the book. As we make new discoveries of ancient ruins, we are able to unravel more of our history and demythologize our past. It is, however, necessary to look at some of the more controversial views that have evolved from our mythological past, to study the concept of our soul's existence. In some cases modern investigators have controversially, and possibly erroneously, turned previously regarded myth and legend into fact. Some of the ideas that have come to the fore as of late have dire consequences for a credible account of our soul's inception and/or existence. Due to our inability to provide credible explanations for how such advanced architecture could have come about way before the Mesopotamians, some theorists have contemplated that they were the result of extraterrestrial visitors. Since my study is about the soul and because of the comprehensive discussion about extraterrestrials, I feel compelled to include something of what is said about their role in the origin of the soul.

As I have now shown, current underwater archaeology indicates there were highly developed civilizations dating back at least 10-12 000 years. Temples of enormous proportions have been found as much as 30-70 meters under the surface of the sea. It is a mystery how these came about considering official history has no record of people with the knowledge or technology to build such temples. Because of the 6 000 year gap in our history, some writers speculate that these yet unexplained civilizations were the result of extraterrestrials or aliens, perhaps even alternative dimensional creatures who came from all sorts of places to inhabit the Earth. Ancient mythology and mysticism also refer to lost civilizations ruled by "gods", one of the modern interpretations for alien ascendance. If this is true, it will greatly influence how we regard our soul nature. If part of our ancestry is of celestial origin, it may mean there are some demigods roaming the Earth.

To illustrate the questionable implications some of these modern-day hypotheses relating to Atlantis and Lemuria bring about, I divide them into three different categories. From a spiritual vantage point and with the origin of our soul nature as my main concern, I find each category not only unbelievable but also unacceptable and sometimes even outright ridiculous. However, since we cannot yet prove that the soul exists, these three theories may well be true. Then again, if we work from the premise that the soul or our soul nature does exist, it is easy to show how these modern-day hypotheses go completely against all rationale and cannot possibly be taken seriously.

Bear in mind that all three of the categories or disciplines I am about to explain refer to exactly the same mythology and period that predates our established and official early urban civilizations. All of them question the lost civilizations before the floods known as Atlantis and Lemuria, but each approach has a different angle.

There is good reason why the lost civilizations are important for our exploration of the existence and nature of the soul. 1. Some people today believe that these lost civilizations consisted of extraterrestrials from other planets and solar systems. 2. Others claim these lost civilizations were the product of alien souls of astral nature seeking earthly experience. 3. A third scenario uses a combination of the two to bring about some hybrid product with ultra supernatural powers. I think it is important enough to know whether this is true, as surely it must affect our lives if we have aliens of any nature for ancestors.

Where and what is the soul in these civilizations? Put differently, did the people of these lost civilizations have souls? It may sound insignificant to ask this question, but if these lost civilizations were considered to be sexually harassed by gods and angels, one should take heed. The same goes for the argument that these gods were extraterrestrials.

From Gods and Monsters

In my opinion, all current hypotheses on the soul based on mythology from the lost cities of Atlantis and Lemuria have either led us in circles or proven to have no foundation for truth. We have no conclusive scientific evidence that we even have souls, yet most of us strongly believe we do. When researching the development of the soul, or anything historical for that matter, it is natural to look for some kind of beginning. Our quest to discover the origin of a soul nature requires an arduous and long journey into the past, where we find the official commencement of civilized history is based on mythology and interpretation.

Interpreting signs, signals and remains from ancient times have left us with countless hypotheses. Some scholars have concluded that some quite colourful and mysterious creatures such as gods, angels, alien beings and other monsters may either be part of our ancestry or may even have been responsible for our existence. Right from Mesopotamian times, documented as the beginning of civilization, we clearly see mythology as being more prevalent than fact. Out of many fragile myths spun around ancient happenings, humans have built more myths until boundaries between legend and fact become blurred, and in many cases, folklore is presented as truth. On close inspection, I see the official version of humankind's recorded history as an elaborate example of how vibrant the imagination can be.

Interestingly, or should I say "dangerously", we have a host of material written by influential researchers who base their theories about our beginnings on myths and conjecture, leaving us with stories resembling science fiction. In the first part of this book, I will be exploring three such beliefs and look at their origins and the consequence they hold for the human soul. One of these beliefs originates from the ancient civilizations of Mesopotamia and describes earlier civilization than themselves as being ruled by gods and angels. In Chapter 2, I show how prominent researchers have taken this concept further by explaining ancient mythological gods and angels as extraterrestrial beings with biological bodies much like our own. They maintain that with extremely advanced technology extraterrestrials made their way to Earth in space ships and invaded our planet. This constitutes the second belief, which I will discuss further in Chapter 3, where we have a substantial number of advocates who maintain that humans have seemingly visible but indestructible alien souls with the ability to transmigrate or reincarnate from one life to another. With something of a spiritual, albeit parasitic nature, they somehow gain supernatural access to embryos in the wombs of unsuspecting mothers. Both these notions are presented with the third possibility in Chapter 4 with the assumption that these souls could be alien invaders from alternative physical destinations or neighbouring planets. These creatures are so advanced that they are capable of transforming from physical to spiritual nature by transmigrating from one dimension to the next at will. They are called the Reptilian Anunnaki. Ghastly! It sounds more like a horror story better left to Hollywood.

The original beliefs or myths from Egypt and Mesopotamia have all played a major role in shaping many of our modern-day hypotheses. My concern is that these conclusions are based on monstrous suppositions, leaving us with alien souls on the one hand and extraterrestrials on the other. Unfortunately, our knowledge of the development of the soul is so limited we have no choice but to scratch around in hearsay, myth and legend and try to separate fact from fiction to gain some perspective. However, on our side are sophisticated research methods making it more possible than ever before to eliminate fiction by taking a scientific approach rather than relying on a religious approach based on faith or dogma. What we have so far are ancient myths and stories accompanied by numerous different interpretations, including incredible postulations and sensational versions, which seem to get more outrageous as time goes by. Let us begin by looking at some of the more extreme conclusions and continue separating fantasy from fact.

In essence, the Egyptian sentiment on the soul implies that we as humans have a spiritual part to our nature, which is immaterial and immortal. Once the final breath leaves the material body, it dies. The soul (called *Ka*) resembles the breath in ancient Egyptian tradition. They also believed the material body to be instrumental in the after-life, hence the practice of embalming and mummifying the corpse to carry the being into the underworld. Without breath or "ka" there could be no life. The Egyptians were among the first to regard at least certain humans such as the kings and queens to be immortal due to a life-giving soul. This part of Egyptian doctrine is factual and still quite credible. Spectacular theories comprising alien extra-dimensional soul beings that are able to take command of our human embryos are not.

Before embarking on a journey to discover whether we do have souls, it is appropriate to look at how far current research and ancient mythology has taken us. As far as I am concerned, there is very little credibility in all three scenarios incorporating supernatural creatures as the shapers of our spiritual existence. The following illustrations will assist in understanding, at a glance, some of the theories that have evolved from our mythological past.

Chapter 2

1 – Extraterrestrial Ancestry

Until very recently, we used archaeological findings from the southern parts of Mesopotamia as evidence of our origins as an advanced urban civilization. Mesopotamia is an area between the Tigris and Euphrates rivers stretching between the Mediterranean Sea and the Persian Gulf. The early settlers in this region inhabited the area more than 7 000 years ago and were known as the Ubaidians. The people who later became known as Sumerians may have comprised of the people from Mesopotamia who interbred with other newcomers from the north-eastern mountainous area (modern-day Iran). Ancient Sumer appears to have consisted of a confederation of cities, all with similar patterns of culture. Over a period of about 2-3 000 years, these ancient people developed language, writing systems and advanced religious, political and legal systems. They dominated the area for several millennia before declining in power and becoming defenceless against foreign invasion around the 23rd century BC.

The earliest known logographic writing from this area dates from around 5 000 years ago, making it the oldest form of written texts ever found. However, archaeologists in China have discovered ancient signs estimated at 8 600 years old, carved on shells and predating the Mesopotamian discoveries by about 2 000 years. Explorers and archaeologists have unearthed important ruins from this time confirming a number of biblical sites. At some of these sites an abundance of inscribed clay tablets, with immense historic significance about the cognitive state at the time, were discovered. Many have been translated and found to be literary and administrative texts with some containing myths and history detailing achievements. For the first time scholars also uncovered texts containing mythical stories that clearly resemble subsequent biblical traditions.

Mesopotamia was the centre of Sumerian, Babylonian and Assyrian civilizations. It seems the inhabitants of the ancient Sumaria and later Babylon believed the universe was ruled by a pantheon of beings that possessed superhuman powers. These "gods" were immortal and each individual god was in charge of a particular component of the various elements of the universe. The Babylonians believed each component of the cosmos, like the realms of heaven (the stellar bodies of the sun, moon, and planets), was ruled by higher-ranking gods than that of Earth (the sea, air, rivers, mountains and cities), all with supernatural and superhuman powers. Although these beings were invisible to the human eye, they were thought to be human in form. Each person had their own personal god or angel to pray to for their salvation.

The Sumerians had a strong concept of the four major elements being fire, earth, air, and water. This is seen in their worship of the leading Sumerian deities such as Enlil, the chief god of the air, Enki, who was associated with water and Ninhursag, who was associated with earth. This is the first time we see the concept of the universe consisting of these four basic components. The early Greek philosophers and scientists also employed this concept. Modern occultists still use this analogy, adding a fifth called the akasha, also known as quintessence, which is an invisible higher realm binding the first four elements together. At the end of the 20th century, scientists started questioning whether this quintessence may perhaps be the missing 95% of the universe's dark energy that has been eluding them for some time. (See Chapter 13 & 18)

Inscriptions on some of the tablets refer to some of the elite gods known as the Anunnaki or sometimes the Igigi. Anu the Sublime, who lived in heaven, was King of the Anunnaki and Bel was considered lord of heaven and earth. Subject to the higher divinity of Gods such as the Anunnaki or Igigi were other lower ranks of gods such as the moon god, the sun god, the war god, the storm god, and even a god of plagues and ruler of the underworld.

During the rule of Hammurabi (king of Babylon) in the 18th century BC, the first set of written laws for the conduct of individuals and society was codified. The Code of Hammurabi inscribed on a pillar is one of the most significant ancient legal documents ever uncovered. The code describes how, during the reign of Marduk, the Babylonian city god became an accepted leader in the pantheon of deities. They assigned the destiny of the land and earthly man to the principal god Marduk and by doing such a fine job he was esteemed among the Igigi.

Many different versions of names and ruling arrangements of gods exist in the translated texts, but there is a common theme in most emphasising a patriarchal society and a nation with a warrior-like nature filled with battles and wars. One of the stories describes how the gods rebelled against working so hard on the all-important irrigation canals. Their request to the authorities to create humans to assist them was apparently granted.

Although nothing is actually known about these ancient settlers' view on the possible existence of a soul, I start my search here for its setting of myth, ritual and magic still prevails in some of our doctrines today. From this era comes the first tangible documentation of a belief in some form of divine realm of spiritual nature that supersedes our own. Although ritualistic burials existed from much earlier times, indicating some sort of perception of a continuing soul, this is the first time we see humans document their understanding of supernatural deities outside their tangible realm of existence. This period of humanity marks a sentient exploration for something more spiritual in nature, a search that was perhaps not consciously acknowledged before. Mythology inherited from these Mesopotamian inhabitants has fuelled many belief systems and supernatural dogma still prevalent today. One of these is the current belief that at least some of these gods referred to in these ancient texts could have been extraterrestrials from other parts of our universe.

Sumerian mythology refers to "gods from the heavens" that ruled the earth, and called them Anunnaki. These invading creatures are later equated with the Nephilim in Hebrew Scriptures as seen in the Old Testament of the Bible. Genesis 6:4 states: "*The Nephilim were on the earth in those days, and also afterward, when the sons of God came in to the daughters of men, and they bore children to them. These were the mighty men that were of old, the men of renown.*"

This scripture is extremely controversial in two aspects. Firstly, it talks of an unknown rebellious bunch called "sons of God" taking wives. They were interpreted as sons of the Creator God (also known as Yahweh) when the Bible text was originally translated. Secondly, the offspring called "Nephilim" are sometimes referred to as "giants". The "sons of God" in this context must actually read "sons of gods". God in this context is highly unlikely to be Yahweh, but rather the Hebrew **ĕlôhîym**, which is plural and can intermittently be applied by way of a courteous expression of esteem to entities with authority like magistrates, judges or even angels.

The contentious part comes in when modern theorists use the two interchangeably and say that the sons of gods referred to and the Nephilim or Anunnaki are actually one and the same. However, the scripture infers that the sons of gods having sex with earthling women would produce an offspring called Nephilim, who could have been giants due to their size or their ranking in the community. Anyway, the possible crossbreeding between divinity and humans is something that has plagued researchers for some time.

By taking the ancient Mesopotamian mythology and later Biblical text literally, some of the contemporary writers have determined that references to "gods from the heavens" and "sons of gods" can be interpreted as invading extraterrestrials from alternative destinations in the universe. In recent hypotheses presented by writers like Erich Von Daniken and Zechariah Sitchin, they make a distinction between the gods and their sons as spiritual entities or royalty with titles, and claim they were rather corporeal extraterrestrial beings who came to Earth in space ships. Granted, these beings must have had unbelievably sophisticated technological abilities to make such journeys possible. The Nephilim or Anunnaki were so advanced they could travel through space from distant planets to reach Earth, a feat still entirely impossible even with our most recent and modern human space technology or envisioned near future science. The arrival of these creatures supposedly explains the sudden evolutionary quantum leap humankind underwent some 12 000 years ago as explained in the previous chapter.

The story goes that inferior earthlings were invaded by what the ancient Sumerians and Hebrew writers referred to as gods. However, the question is whether these gods, their sons and offspring referred to in these ancient texts were divine beings, merely earthlings with lofty titles, or as now suggested, perhaps extraterrestrials from outer space? Were they really extraterrestrials with such phenomenal knowledge, technology and ability that earthlings likened them to heavenly gods and even perhaps angels? Could such extraterrestrials be responsible for the evolutionary quantum leap that took place in the history of humankind some time before the Mesopotamian era? Either way we end up with powerful incorporeal gods capable of controlling the elements on the one hand, or corporeal rulers also sometimes referred to as gods who in the context of this chapter are interpreted as advanced invading extraterrestrial beings.

To find out how the modern-day theories of extraterrestrial visitation took place we have to revive the story of Lemuria. The underlying argument in favour of extraterrestrial interpretation is based on the quantum leap seen in ancient mythology. These progresses, combined with art and elaborate architecture, as well as newly found oceanic archaeological findings dating back some 12 000 years, do not seem to fit our current historic timeline. We still do not understand how it was possible for humans to create such spectacular monuments when we cannot yet connect our nomadic hunter-gatherer existence with the

Neolithic revolution that started at the very earliest around 10 000 BC in Mesopotamia at least 2 000 years later. This has led to a search for alternative explanations, including one in which we were visited and possibly even invaded by superior extraterrestrial races now arguably said to be the Nephilim or Anunnaki. These extraordinary creatures are said to be the designers and early rulers from the legendary place called Lemuria.

At the beginning of the 20th century, James Churchward wrote extensively about a lost civilization called Lemuria and published well-known books on the topic, including *The Lost Continent of Mu* in 1926 and *The Sacred Symbols of Mu* in 1933. He is one of the first researchers to present Lemuria as a lost civilization similar to that of Atlantis and was adamant that a search for evidence of its existence would be successful. (Remember this was written long before the current discoveries of lost temples under the sea). He does not go as far as saying these lost civilizations were of extraterrestrial origin, but claims they were as superior to our own Mesopotamian origins – even though he dated the Lemurian civilization much further back than modern anthropology has determined our evolutionary timeline.

Churchward spent his life searching for the continent of Lemuria, travelling all over from the vaults of an Indian temple to Australia, Siberia and the South Seas, visiting remote monasteries in Asia and studying ancient records of the "Motherland", Mu or Lemuria dating back between 12 000 and 70 000 years. These records tell of a vast civilization which emerged, flourished and decayed. In his book *The Children of Mu*, he describes how various racial types, including blue-eyed blonds, peopled the world. Churchward claims that Lemuria was destroyed about 12 000 years ago.

Ian Lawton, author of *Genesis Unveiled*, alleges that Churchward's works are not referenced properly and are all supposedly based on two sets of ancient tablets. Lawson suggests that the tablets Churchward used to base his studies on were composed in Burma or even in Mu itself. They were shown to Churchward by a temple priest in India, after which the two of them proceeded to decipher them together, Churchward having studied the "dead language" with the priest the two preceding years. From Churchward's interpretations, the tablets contain little more than a basic esoteric view of world origins common to most Eastern and in fact Western traditions. Lawson claims that the tablets contain no obvious references to Lemuria itself, that the interpretation cannot be corroborated and that they contain nothing particularly revelatory.

Another set of tablets used by Churchward and incorporated into a second edition of *The Lost Continent* in 1931, are purported by Lawton not to date back further than 12 000 years despite what Churchward claims. In essence, Lawton believes Churchward relied heavily on Le Plongeon's distorted interpretations of the Troano Codex and other Mayan inscriptions and finally, despite his apparently more sensible time-scales, Lawton quotes Churchward as admitting to being a creationist, casting further doubt on the reliability of his work. Despite all this scepticism, Lawton accepts Churchward's records regarding the location of a number of megalithic stone structures that are still standing on the Pacific islands. Some of these include enormous stone statues, pyramids, an open-air temple, huge terraces and vast columns.

A controversial researcher, Erich Von Daniken was widely recognised in the 1970's with the publication of some 20 books, *Chariots of the Gods* being one of his bestsellers. Von Daniken is passionate about the reality of extraterrestrial visits and has a philosophy he calls Palaeo-seti (palaeo meaning old or ancient from the Greek word "palaios" and "seti" meaning search for extraterrestrial intelligence) (*The Return of the Gods*). Von Daniken makes no bones about the fact that he believes real extraterrestrials visited Earth thousands of years ago. Although he agrees his theory cannot yet be proven, he offers a structure of ideas as a helpful foundation to explain the impossible, attempting to clear up complications caused by religious views and beliefs and open a new path of thinking on these matters.

Referring to various different sources and scriptures to support his ideas, he believes that most references to "sons of God" that came from heaven or the sky can be more correctly translated by Hebrew specialists to mean extraterrestrials travelling on Unidentified Flying Objects (UFO's), rather than interpretations like "giants", "children of God", "fallen angels", or even "renegade spirit beings". Von Daniken is adamant that crossbreeding with the extraterrestrials caused humans to receive genetic messages inconsistent with their human ones. He thinks the human experiment has somehow gone awry.

Von Daniken presents another point of view, which maintains that when the giant mother-spaceship of the extraterrestrials entered our solar system, the passengers already knew about the third planet, which was Earth. They also knew that only on this blue planet could life exist. Earth had many different life forms including our primitive ancestors who, even though they were slow-witted, represented the highest form of life at that time. After successfully genetically altering some of these beings (who then became the first Homo sapiens), they decided to leave Earth.

In *The Return of the Gods* (p. 29-31), Von Daniken relates his interpretations of the biblical Eve being visited by an UFO. He says one can read about these flying chariots during Adam's time in some apocryphal text written by unknown hands in 730 BC, but which may be based on handwritten documents from earlier times. (Von Daniken refers to the apocryphal text "*The Life of Adam and Eve*" published in the *Daily Telegraph*, 7 April 1993.) It tells the story of Eve looking towards heaven and seeing a chariot of lights pulled by four gleaming eagles of magnificent beauty. Von Daniken wonders if Eve was the first person to witness a UFO. Von Daniken then suggests that since the text mentions the Lord mounting the chariot, the Lord who created Adam and Eve climbed aboard this UFO.

In *The Return of the Gods* he also relates his interpretation of sons of gods and renegade angels. The Jewish legends also allege that after Eve's seduction two separate races evolved from the descendants of

Cain and Abel. The descendants of Cain behaved like animals. Here the Jewish legend seems to elaborate on the story told in Genesis 6 about the Nephilim and sons of God creating havoc amongst the earthlings. With the collapse of morality and sexual antics in Sodom, "fallen angels" descended from the heavens and took "human wives". Their offspring, also said to be giants, are the same Nephilim or Anunnaki represented in other contemporary writers' works. The giants, who were of mighty girth, occupied themselves with robbing, plundering and even spilling blood. They also produced offspring and as many as six of them were delivered simultaneously at birth. The Jewish legends identify different giant races such as the Emites or Frighters, the Rephites or Gigantics, the Giborim or Mighties, the Samsunites or Sly Ones, the Avids or Wrong Ones, and finally the Nephilim or Spoilers.

Von Daniken speculates from here onwards, based on hearsay and texts by unknown writers. He maintains that the crossbreeding of extraterrestrials with people produced creatures that were entirely inconsistent with the planned race of Homo sapiens. This is said to be the original sin of mythology. Human beings were now inheriting the wrong genetic messages, which made the Lord grieve and repent that he had made man on the Earth (Genesis 6:6).

Perhaps one of the questions still to be answered is why these early scriptures, dating from before the Egyptian epoch, did not make any mention of the soul. Von Daniken points out that the gods created men in their own image. There is no mention of the extraterrestrials having souls or whether they genetically engineered humans with souls. I would imagine that if a race were capable of genetically engineering or creating such despicable offspring, the creators would be careful how they produce their successors so they do not end up as advanced or powerful as themselves. I would imagine these humans must have been created without souls if they were a genetically engineered slave race. I am certain souls would have no purpose for their creators, since that would give them too much power beyond death. However, it could be that these humans did have souls since Adam was created with a soul, and that the extraterrestrials did not. It is horrifying to think what features these offspring may or may not have ended up with, whether they were born with or without souls. Imagine if not everybody can lay claim to such a reassuring life-saving feature as the soul because of an extraterrestrial ancestry.

Von Daniken's assumptions go off the chart by inferring that the "highest" being – the commander of the spaceship – possessed better maps than his renegade crew. The "highest" must somehow interrupt the "human being" experiment and begin afresh. The renegade angels probably possessed powerful weapons, which they could hide away in caves or hold inside buildings, and there was no possibility of hunting the evil ones individually. So a giant mother-spaceship containing extraterrestrials cruised into our solar system. Those aboard had known about our planet for some time. The extraterrestrials are said to have taken one of the earthling creatures and altered it genetically. At some point, a group of extraterrestrials found their experiment had succeeded and this is how Homo sapiens came into being. After some time the extraterrestrials left planet Earth to this new being. To multiply, a female was needed and lo and behold, Eve is created!

Von Daniken elaborates with more quotes and stories with similar interpretations, but I will not go into any more detail since the story just gets more unbelievable. One area perhaps worth looking at is the *"Book of Enoch"* which asserts that "real" angels mutinied on Earth. The ancient script, which is named after the Hebrew patriarch Enoch and first descendant of Cain (Adam's son), is a composite of sections written by different authors at various times during the 2nd and 1st centuries BC. In *The Gods were Astronauts,* Von Daniken mentions the apocryphal script representing the corroboration of his palaeo-seti philosophy. He accounts how Enoch, like the Old Testament, gives its own version of what happens when angels mutiny. The Book of Enoch (6:1-6) says the children of men multiplied and lovely and loving daughters were born as a result. The angels and/or the sons of heaven saw them and lusted after them, telling themselves to take these earthling lovelies as wives and let them bear children. Semiaza, their leader, asked them not to go through with this for he would be held responsible for such transgression, but eventually they all swore an oath and bound themselves to it. Von Daniken correctly points out how this could be construed as mutiny. So, all of them took wives unto themselves and carried out impure acts with the earthling women. Somehow, when the extraterrestrials ran out of food they turned against them and devoured them. They began to devour birds, wild animals, creeping creatures and fish as well as one another. Then the Earth cried out aloud against these fiends. The good angels who had not become involved in the mutiny, apparently observed all this from above.

An important academic, one of the few scholars able to read and understand the Sumerian language, is Zecharia Sitchin who was born in Russia and raised in Palestine. Sitchin is said to have sound knowledge of Hebrew and many European languages, studying the Old Testament as well as the history and archaeology of the Near East. He is the author of books such as *Genesis Revisited, The 12th Planet, When Time Began* and *The Earth Chronicles*, which were written from translations taken directly from clay tablets from ancient civilizations.

Sitchin puts forward the possibility that the extraterrestrials came from a planet called Nibiru, a 10th planet, which he says is part of our own solar system. The other nine are commonly divided into two groups: the inner planets (Mercury, Venus, Earth and Mars) and the outer planets (Jupiter, Saturn, Uranus, Neptune and Pluto). Sitchin surmises that after a clash between two planets the inhabitants from both Nibiru and Mars came to live on Earth. However, Sitchin is only elaborating on what seems to be part of the earliest recorded historical myths.

As with Von Daniken, the main theme of Sitchin's work is based on the writings of ancient Sumerians. He tells a similar story about Sumeria – how it appeared suddenly and seemingly out of nowhere some 6

000 years ago. He also tells us how it is credited with virtually all the firsts of a high civilization such as inventions and innovations, concepts and beliefs. These innovations form the foundation of our own Western culture and those of all other civilizations and cultures throughout the Earth, including the wheel, animal drawn vehicles, boats, ships, the kiln and brick, high-rise buildings, writing and schools, a legal system, dance, music, medicine, religion and more. He emphasises how, in particular, advanced astronomy was widely documented showing concepts like the celestial sphere, the horizon and zenith, the circle's division into 360 degrees, a celestial band in which the planets orbit the sun, the grouping and naming of stars into constellations, the creation of the zodiac, and applying the number 12 to the zodiac as well as devising a calendar that forms the basis of calendars to this day.

According to Sitchin, the reason we know this is that the Sumerians recorded all these details on clay tablets. They drew their illustrations in reverse on cylindrical seals, to appear as a positive when rolled on wet clay. In the ruins of Sumerian cities excavated by archaeologists in the past century and a half, hundreds, if not thousands of the texts and illustrations that were found dealt with astronomy. Among them are lists of stars and constellations documented in their correct heavenly locations and manuals for observing the rising and setting of stars and planets. There are texts among the unearthed tablets that list the planets orbiting the sun in their correct order; one text even shows the distances between the planets. Featured prominently in their cosmogony was a cataclysmic event they called the Celestial Battle. It was an event to which references were made in miscellaneous Sumerian texts, hymns and proverbs similar to some of the Old Testament books of the Bible. It was written in great detail in a long text made up of seven tablets. It deals with the formation of the solar system prior to a Celestial Battle and with the event of an awesome collision resulting in a 10th planet in our solar system called Nibiru.

Homo angelicansis

Sitchin tells us the Anunnaki came to Earth from the 12th member of the solar system or, technically speaking, the 10th planet. It is from this planet the Sumerians claimed the Anunnaki originated. The advanced knowledge we find in Sumerian texts is knowledge passed on to us by the Anunnaki who had come from Nibiru. Sitchin explains that the civilization from Nibiru was very advanced and that they were already travelling in space 445 000 years ago when they came to Earth. Their vast elliptical orbit made a loop around our solar system, acting as a moving observatory from which the Anunnaki could investigate all our planets. This, Sitchin says, is the reason why our current discoveries are similar to what was already known in Sumerian times.

The Sumerian texts answer the question why the Anunnaki wanted to come to Earth every 3 600 years. On planet Nibiru the Anunnaki were facing ecological deterioration (just as Earth is) and this made life increasingly impossible. There was a need to protect their decreasing atmosphere and the only solution seemed to be suspending gold particles above it as a shield. The Anunnaki discovered gold on our planet and started missions to mine the precious metal. Initially they tried to obtain it from the waters of the Persian Gulf, which failed, so they started mining in south-eastern Africa. Some 300 000 years ago, the Anunnaki who had been assigned to mine the gold in Africa, supposedly mutinied. Obviously not happy with all the hard work, the chief scientist and the chief medical officer of the Anunnaki used genetic manipulation and in-vitro fertilization to create the first Homo sapiens to serve as "primitive workers" or a form of slave miners.

Sitchin says the discovery of Nibiru would confirm that there are intelligent beings out there who, almost half a million years ago, could travel in space. He points out how photographs of the surface of Mars show similar structures to those on Earth attributed to the Anunnaki, making some people believe that Mars was once used as a space base and even that this space base has been reactivated. In recent years we have seen many writings about Nibiru, which is also known as Planet X. Some speculators believe the Anunnaki from Nibiru are coming back to Earth soon. In fact, Planet X is said to have a 3 600 year elliptical orbit around our sun. According to certain researchers, we therefore should have seen these extraterrestrials return in May or June of 2003 (we must have missed the landing!).

At the end of July 2005, it was announced that Mike Brown, Chad Trujillo and David Rabinowitz from Palomar Observatory near San Diego, California discovered an unconfirmed 10th planet called 2003-UB313 orbiting our sun. The planet is apparently covered in methane ice and at nearly 15 billion kilometres from Earth, it is the most distant planet in our solar system. Although the planet was first seen in October 2003, it was too far away and could only be verified on January 8, 2005. Astronomers have also discovered a moon (named Gabriella) of almost a tenth the size orbiting 2003-UB313. Even though the scientists have proposed the name Xena for the newly found planet, it remains to be seen if it would end up as Nibiru and whether it meets the criteria set by Sitchin.

From the mountains to alluvial plains

How is it possible we can be taken in by such fantastical stories? Without hard and fast facts, the countless different interpretations of theologians, archaeologists, researchers and the like remain pure speculation. Even when discoveries of ancient written texts are made, the differing interpretations carry many contradictions and controversy. None of the various researchers agree with each other, even though many highly developed thinkers have pondered long and hard over the texts available to them. Although the extraterrestrial conspiracy cannot be taken lightly, neither can it be taken seriously. There simply is no concrete scientific proof, merely a collection of prevailing fabulous myths, legends and personal interpretations. Von Daniken admits that the human imagination is a powerful force in fuelling myths, realising that folk tales depend on fabulous stories. Ultimately, thousand year-old texts can only describe something that could have happened, they cannot tell us if it actually did.

I find it difficult if not impossible to think that these "sons of gods" or "gods from heaven" were extraterrestrials and would also forfeit the idea that they were offspring of a real divine and Almighty God as represented in Hebrew legend. It makes more sense that they were the royalty and rulers of the time who were described as gods and the likes.

The hypothesis that these beings were mere rulers and earthly royalty is clearly described in the works of David Rohl, author of the best-selling history books *A Test of Time: The Bible – From Myth to History* (1995) and *Legend: The Genesis of Civilisation* (1998), which have since formed the basis of the documentary programmes *In Search of Eden*, *The Egyptian Genesis* and in particular *The Lost Testament* (2002). Rohl holds a degree in Ancient History and Egyptology (University College London) and was awarded the prestigious W. F. Masom History Research Scholarship by the University of London. His views offer an alternative explanation of Yahwism, casting new light on some of the Biblical doctrines.

Rohl maintains the name Anunnaki can also be understood as "the ones of heaven and earth" – in other words those who came from heaven (*an*) to settle on earth (*ki*). For him heaven and earth means these people came from the *mountains to the alluvial plains*. They were mortal men who became gods. Rohl purports they were earthbound gods of flesh and blood, who passed from the mortal world to govern the underworld of the dead. They did not dwell amongst the Igigi gods of heaven. The Egyptians also knew them as the mythological founders of civilisation, the revered Shebtiu or the Senior Ones.

Throughout the history of human civilization heaven has been interpreted to be synonymous with high grounds or mountainous areas. In all of human existence, we have seen a struggle for survival and dominance beset with war. Perhaps there was great significance in controlling the high ground, particularly during times of floods and catastrophe. That settles that one for me! "Gods" are earthlings with titles.

Chapter 3

2 – Invasion by Alien Souls

In the previous chapter we looked at how the mysticism of the past was used as a basis for the belief in extraterrestrials invading Earth and becoming the genetic masterminds and creators of our ancestors. Our ineptitude at combining new data with ancient doctrines has resulted in a belief that our ancestors were partly created or genetically engineered by extraterrestrials. That possibility does not take into account whether these extraterrestrials created humans with souls or not. If they did possess the technology to apply genetic manipulation, it was most likely on a biological level similar to our own today. The question of spirituality is not considered in this context.

In the next synopsis, we look at a completely new scenario where the extraterrestrials are no longer physical in nature, but now became spiritual aliens called "ka", "atman" or soul. It does not really matter what you call it, it all resembles the same thing: ghostlike creatures from alternative realms capable of transmigrating or reincarnating from one body to the next over lifetimes, only to return to some yet illusive spiritual realm when escaping death of the physical body. It all sounds very convenient. It started in Egypt and ends with the most fabulous delusional imagery imaginable. It does, however, make some good reading.

Apart from the belief in mythical gods by early civilizations, the next period ushers in the belief of developing an individualistic spiritual nature and being. It started with the early Egyptians as a mere breath called "ka" but later, when the nature of "ka" is questioned, it turns into an immortal transmigrating and reincarnating Atman in Eastern religions or soul as it is known in the West. The "alien soul" theory, as I call it, gets bigger and more fantastic as modern humans develop to a point where we have created in our minds a whole realm of ghostly spiritual beings we think we can see and interact with as if a hologram or movie seen through three-dimensional spectacles.

Although most of us are familiar with the concepts of transmigration and reincarnation, the passing or cyclical return of a person's soul at death into a new body or new form of being, we may not be aware that the belief implies our souls are spiritual aliens from alternative realms. Essentially the belief in reincarnation entails spiritual beings, from another realm other than our own visible universe, taking charge of human bodies on Earth as a vehicle for physical or biological life experience. It is speculated that such beings come from alternative vibratory realms, different dimensions or a combination of both or may also simply be from another planet, solar system or galaxy. In this chapter, we deal with spiritual beings called "alien souls" and not physio-biological beings such as the extraterrestrials in the previous chapter.

Probably the oldest and foremost religious doctrine on the soul incorporating transmigration and reincarnation emanated from Egypt. The Egyptians embalmed their dead to preserve the body so it might accompany the "ka", the animating force which today means soul, into the next world. We are not sure what went before, but we pick up the trail at a point where there is already an understanding that the soul exists in the form of something like breath, which is a life-giving entity and can exist after biological death. Egyptian eschatology is well established by the time of the pharaohs, and their belief in afterlife and transmigration of the soul takes centre stage. Although some form of belief in spirituality and afterlife dates back as far as 30 to 40 000 years ago, it is not until now that it is given a name and described in much detail in text, burials and art. It is at this time in history that we are dealing with more than simply physical beings; we are dealing with a belief in a life-giving force of spiritual nature that comes from a spiritual realm and survives death only to return to that same realm as "ka" or soul.

Although Egypt was environmentally hospitable in the seventh millennium BC, and evidence of settlements from that time is found in the low desert areas of southern or Upper Egypt, similar remains have been discovered at Nubian sites in modern Sudan. Egypt only started to become a formidable force in 3 000-2 500 BC. Archaeological sources indicate the emergence of a dominant political force, around the late Gerzean period (about 3 000 BC), that was to become the consolidating element in the first united kingdom of ancient Egypt. Here we see the earliest known hieroglyphic writing and the names of early rulers beginning to appear on monuments.

Settlements developed along the Nile valley from the fourth millennium BC. They had irrigation systems for their crops, tombs and various crafts like pottery. Amongst the elaborate rituals regarding the dead is a *Book of the Dead*, a large collection of funerary texts of various dates containing magical formulae, hymns and prayers believed by the ancient Egyptians to guide and protect the soul (ka) in its journey into the region of the dead. Knowledge of these texts apparently enabled the soul to ward off demons attempting to impede its progress, and to pass the tests set by the 42 judges in the hall of Osiris, god of the underworld. These texts also indicated that happiness in the afterlife was dependent on the deceased having led a virtuous life on Earth. The earliest known religious (funerary) texts were found cut in hieroglyphs on the walls inside the pyramids of the fifth and sixth Dynasty kings of the Old Kingdom. These became known as the Pyramid Texts. Burying the dead was of great importance and Egyptian funerary rituals and equipment became the most luxurious the world has ever known. The departed are routinely called the "living", the coffin is the "chest of the living", and the tomb the "lord of life". The Egyptians considered the dead free to revisit the earthly tomb, and in making the journey back and forth, they had the power of transforming themselves at will into various animal shapes.

It is not merely the disembodied spirit or the soul, as we understand it, that continues to live, but the soul with certain bodily organs and functions suited to the conditions of the new life. In the elaborate anthropology which underlies Egyptian eschatology, which we find hard to understand, several constituents of the human person are distinguished (the most important being "ka"). These constituents are believed by the Egyptians to be restored to those who pass judgment after death. However, the new life that begins after a favourable judgment is not at first any better or more spiritual than life on Earth. If the judges decided the deceased had been a sinner, the "ka" would be condemned to hunger and thirst or even capital punishment. If the decision was favourable, the "ka" went to the heavenly realm of the fields of Yaru, where grain grew to more than twice the height of a man and existence was a glorified version of life on Earth. The "ka", however, could not exist without the body and so every effort had to be made to preserve the corpse. Isis, who mummified her husband Osiris, supposedly started mummification and embalming. People believed in transmigration of the soul long before ancient Egypt, but this is the first real record of the extent of their belief.

The concept of a soul developed with further acknowledgment that the soul departed from the body at death and entered a new body at the next birth. The Egyptians thought that the soul could leave the body during sleep. This soul was pictured as a vapour that entered and left through the nostrils and mouth. Later the notion grew that the soul transmigrated to an infant from the dead person's family. This helped to explain family resemblances. "Second death" and other expressions deriving from the hieroglyphics might seem to suggest annihilation; but it is sufficiently clear from the evidence as a whole that continued existence in a condition of darkness and misery was believed to be their lot.

In Babylon (the principal city in Mesopotamia) as in Egypt, the Babylonians considered the life of the king to be directly related to the life of the land, its crops, herds and its people. Differing from Egyptian belief, the king was not himself considered divine, but was highly regarded as a representative of God. Ancient Babylonian and Assyrian religion never attained any higher degree of development. Yet they believed in the hereafter. They believed a kind of semi-material ghostly shade or double (*ekimmu*) survives the death of the body, and when the body is buried, the ghost descends to the underworld to join the company of the departed. The location of the region of the dead is controversial, but the suggestion is they believed in resurrection or some form of return to Earth from the dead, which is sometimes concluded from Babylonian belief in the "waters of life" and from references to Marduk as "one who brings the dead to life".

Indian Hinduism has some similarities. In 1920, the excavation of two ancient and forgotten cities of the Punjab gave us evidence of India's early history. Before this, the only records were the Vedic sacred texts with information from about 1 500 BC when the Vedic Aryan tribes entered North-West India. Books about Indian religion written before 1920 rely heavily on Vedic texts. Much speculation exists, but the Indus Valley was already highly developed in the third millennium BC with town planning, brick structures and drainage. The civilization seems to have been at its peak between about 2 500 and 1 800 BC, after which there is little archaeological evidence of settlement in the cities, suggesting these were abandoned. The reasons for the decline are uncertain. Environmental changes, earthquakes or invasions are suggested as causes for this. Needless to say, the Indus mixed with the Aryans and became Vedic India.

Migrant Indo-European tribes arrived in Dravidian India in about the middle of the third millennium BC. These tribes came through the mountain passes along the North Western frontier and gradually settled in most of the territory north of the Vindhya Range and west of the River Yamuna. There is no evidence that these population changes were in the form of invasions. Political details around this time are sparse, but a collection of sacred writings called the Veda dating from about 1 200 BC, contains substantial information on their culture, social practices and religious beliefs.

Many other cultures also believed in various forms of transmigration, similar to the Egyptians. The general belief is that the body is inhabited by a single soul or vital essence, which separates from the body at death and during sleep, passing in and out of the mouth or nostrils. While separated from the body after death, the soul seeks to inhabit a new body, and if need be will enter the body of an animal or some other lower form of life. Reincarnation was also believed to occur by transmigration of a dead person's soul to the body of an infant of the same family.

In eastern religious thought, the belief in transmigration does not seem to be part of the most ancient religious beliefs of the Aryan conquerors of India. It first appears in the Upanishads, which is part of the Veda, the most ancient sacred literature of Hinduism written in the Sanskrit language. Initially Indian religious belief included immortality, which is a simpler doctrine than the Brahmin and Buddhist forms that succeeded it. Later the purification of the soul became dependent on a series of ritualistic beliefs.

The Upanishads identify the "atman", which is the Hindu term for the breath or soul, and the principle of life with the divine Brahman, which means the Absolute, the omnipresent Reality considered as pervading the universe. The crucial part of this belief revolves around rebirth and the identity of atman and Brahman. Life continues endlessly due to rebirth ("samsara") of the soul (atman) by being able to reincarnate from one body after dying only to be replaced by another body. This next body is determined by the actions of previous lives based on the concept of "karma". Karma is said to be the sum total and virtue of one's actions that are attached to the soul as it transmigrates. Positive karma gives rise to happier and more favourable rebirths, whereas negative karma brings on contemptible rebirths.

After death, there are many heavens or hells for reward and punishment of all degrees of virtue, but these conditions can be seen as preliminaries to further rebirths in higher or lower forms. Salvation, to which man may ultimately look forward, is achieved only when you are able to escape this endless cycle of rebirths. The atman or individual soul is believed to be identical with Brahman, the universal soul. Those few who have acquired a perfect knowledge of Brahman are able to make such escape possible. In Hindu philosophy, the term "atman" also designates the true essence of anything including the universe. The atman or soul in this context is the immortal substance that is able to transmigrate from body to body. It is also regarded as the principal informant of all activities, identity and consciousness. The body houses the atman and in this way separates it from Brahman, which is beyond all material forms. Through meditation on Brahman, desires and ignorance can be overcome and the atman can be set free to be one with Brahman. Brahman itself may be characterized as "true being", "consciousness", or "bliss", but in the final analysis, it is perceived to be beyond any language. According to this doctrine the soul is captured in the human body during earthly incarnations until it achieves purification and knowledge and can eventually merge with ultimate reality, where it is freed from the cycle of birth and rebirth (known as "moksha"), meaning liberation.

The ancient Greeks around 500 BC regarded transmigration as a doctrine closely associated with the followers of the philosopher and mathematician Pythagoras. According to Pythagorean teaching, the soul is immortal and survives bodily death. After a series of rebirths in other bodies (each rebirth following a period of purification in the underworld), the soul eventually becomes free from the cycle of reincarnations. Plato maintained that the soul is eternal, pre-existent, and wholly spiritual. After entering the body, it tends to become impure through association with bodily passions. It nevertheless retains a minimal knowledge of former existences. Delivery from the body occurs only after the soul passes through a series of transmigrations. If the soul has had a good character in its several existences, it is allowed to return to a state of pure being. If, however, its character has continually deteriorated in its transmigrations, it ends in Tartarus, the place of eternal damnation.

The idea of transmigration was never adopted into Orthodox Judaism or Christianity. Among Jews, only the mystical Cabbalists adopted it as part of their system of philosophy. The Gnostics and the Manichaeans also believed in transmigration, but early Christians who adopted Gnostic and Manichaean doctrines were declared heretics by the Church, when in 533 AD, the Council of Constantinople declared reincarnation a heresy.

The reason reincarnation was repudiated, was because of the eschatological teachings of death and judgment which were established as orthodox Christian doctrine. Simply stated, this doctrine maintains man has just one life in which to merit his eternal reward or damnation. This is basically why Christians are not exposed to or believe in the concept of reincarnation or transmigration. Such a doctrine also strengthened the Church. However, up to this point the widespread belief in transmigration was that a person's soul carries on living in some form or another after physical death.

The format of transmigration and reincarnation did not really change much from then until the late nineteenth century when some occultists started adding more mystique to the already astounding concept. With the Christian Church losing its grip on controversial cultic practices, and with freedom of speech allowing alternative beliefs systems to resurface without persecution, many practitioners started to share their knowledge on the discipline. The introduction of new occult knowledge resulted in a huge revival, allowing many people to practice alternative religious beliefs and even publish their experiences without fear of being burnt at the stake for heresy.

It was at this point that the spiritual entities, which purportedly must still be the original ka, atman or soul, became incorporated in what is known as the development of the root races. These strange and controversial creatures are said to be capable of evolving in a spiritual realm only to then engage human beings before birth. These theories are huge thumb sucks at best, but certainly made an impression on the psyche of humankind since we have succumbed to the delusional belief in a spiritual realm complete with ghostlike creatures, dead relatives, guides, angels and the likes, all of whom visit us.

Madame Elena Petrovna Blavatsky (born Helena Hahn, 1831-1891), the founder of Theosophy and author of *Isis Unveiled* and *The Secret Doctrine* (sometimes referred to as the Bible of the Occult), wrote about transmigration and reincarnation of the soul nature in the late 19th century. *The Secret Doctrine* includes a belief in an Ultimate Reality, as an omnipresent, transcendent principle beyond the reach of thought, similar to Brahman. It also explains Blavatsky's views on the universal law of cycles throughout nature, the likeness of all souls to the Universal Oversoul and the soul's journey in the form of reincarnation, in accordance with "Cyclic and Karmic law".

Blavatsky's work is strongly influenced by eastern religions such as Hinduism and Buddhism. Her eschatology is based on the principles of transmigration and reincarnation of alien souls from another realm wanting to experience physical life in biological human bodies, or shall I say the bodies of human beings. Blavatsky goes into great detail trying to explain where these spiritual beings or souls, our forefathers, originate. No longer are we talking about humans with a soul, but rather a soul taking on a human form. The emphasis is on a soul realm with soul beings wanting earthly experience rather than humans dying and our souls transmigrating. Although we see where her doctrine comes from, let us examine where she (and the Theosophical movement) takes it.

To justify her doctrine that these spiritual beings, over a period of millions of years, acquired the ability to become physical beings, she had to concoct a credible story, hence *The Secret Doctrine*. We start with the basics of a spirit being in a spiritual realm called the etheric realm, first becoming astral beings and eventually entering into the physical phase, which, according to Blavatsky, happened during the Lemurian epoch. Lemuria and Atlantis are respectively the third and fourth epochs of what is known as the Seven Root Races. In applying this cyclic view to the human race, she reveals herself as categorically opposed to evolutionary theory. Blavatsky argues that, until the middle of the third root race, there was no normal sexual reproduction. This forces her into all sorts of pseudoscientific contortions about how these evolutionary changes came about. Worse, she claims that the changeover to normal sexual reproduction in some way forced the creative gods, compelled by karmic law, to incarnate in mindless men.

Blavatsky bases her information on the scriptures of Dzyan, which she claimed the Mahatmas communicated to her. As we do not have access to these scriptures, we cannot verify the credibility of her theory. Having said that, her postulations have a major influence in occult circles and some new age movements, but the entire concept has very little if any merit without access to the documents they refer to.

To lend credibility to this hypothesis, the occultists have created a historic timeline in which these souls developed the ability to gradually shift from spiritual bodies to physical bodies. This underpins the theory of how spiritual development took place in the form of seven root races. The first root race she calls the Adamic race. This race was as ethereal as ours is material. Some occultists say we are an etheric race with astral form. Etheric is a term derived from Theosophical literature to explain the first race of man – souls who were preparing for the descent into a body of matter. They were said to possess only the faculty of hearing and to dwell in protean astral bodies. I will explain the concepts etheric and astral in more detail in a later chapter.

The second root race (Hyperborean or Reptilian) refers to the second continent, housing the second stream of evolution situated in the area we know as Northern Asia. In the distant past, it had no winter and was a land of the gods preceding even Lemuria. In classical mythology, the Hyperborean race was the ancestor of the gigantic Titans, but in the esoteric lore, these mighty beings did not incarnate into physical bodies in the way of modern man. The one-eyed Cyclop giants may be seen as racial memories of these beings that had advanced inner vision now associated with the central Third Eye of clairvoyance. The first and second root races are regarded as purely spiritual rather than physical entities.

The third root race came into being in Lemuria from a continent that emerged more than 18 million years ago and sank about 5 million years ago. According to Blavatsky, the inhabitants were not proper humans yet, but had learnt how to take on a physical nature. They were egg-laying beings with a third eye that gave them psychic powers and allowed them to function without a brain. Originally sexless, the Lemurian race separated into two sexes and developed womb birth. Evolutionists have commented that 18 million years is not a long enough period for man to have developed to this extent physically, astrally and spiritually.

The fourth root race is the Atlantean epoch, but the Atlanteans described by Blavatsky are different from the lost civilizations described earlier, in that she has incorrectly documented these people as living hundreds of thousands or even millions of years ago. Most theorists agree that the Atlanteans were fully developed human beings sharing the basic characteristics of today's Homo sapiens.

The fifth root race or Aryan race is where we are currently. As the sequence of the continents is made to follow the order of evolution of the Races, from the first to the fifth, Europe must be called the fifth great continent. The Secret Doctrine takes no account of islands and peninsulas, nor does it follow the modern geographical distribution of land and sea. Since the day of its earliest teachings and the destruction of the great Atlantis, the face of the Earth has changed more than once. There was a time when the delta of Egypt and Northern Africa belonged to Europe, before the formation of the Straits of Gibraltar and a further upheaval of the continent entirely changed the face of the map of Europe. The last serious change occurred some 12 000 years ago, and was followed by the submersion of Plato's little Atlantic island, which he calls Atlantis after its parent continent.

The next two root races, the sixth and seventh, are still to come and details about them are yet ill-defined.

Since we have no access to the scriptures of Dzyan that Blavatsky referred to, and are faced with numerous distortions, we can remain sceptical of her writings. If the various ancient and sacred texts were to be revealed, it would be interesting to see how closely commentaries on such texts match the originals. However, even if they were to be revealed, we would still question their antiquity and degree of insight. Ultimately, we cannot rule out the possibility that Blavatsky may have simply written her own version of history combining Tibetan, Hindu, Jain and other influences.

Notable in writings from those who include alien souls and extraterrestrials into interpretation with the ancient texts is the idea that we are helpless bystanders, waiting for an outside source to seal our fate. I personally cannot trust any of these assumptions to be correct. There are no proofs to verify the above and the chronology does not correspond with that of modern science. The very latest views of the 21st century are based on the progressive development of the belief in transmigration and reincarnation of the ka, atman or soul.

Michael Newton, counselling psychologist, a certified master hypnotherapist and author of *Journey of Souls* and *Destiny of Souls*, writes extensively on transmigration and spiritual migration from one realm to another by birth into human form. Pythagoras was the first to coin the phrase and describe the philosophy of "Journey" of the soul, which Newton and other current writers use. The "Journey" referred to by Pythagoras most likely represented the transmigration process. Newton has a slightly new angle in that soul beings are able to select a human embryo once it is in the mother's womb, taking possession of it at will to be born into the newly found body that the unsuspecting mother has been carrying. This seems to be a continuation of the early ancient doctrines and that of Blavatsky, the theosophists and occultists, only with an added twist.

Newton's theory and writings are particularly popular, even though he does not say much that has not been said before by the theosophists and occultists. The reason for developing an age regression technique is to take hypnotised subjects beyond their past lives, and into soul experiences between lives. Graphic details are given about how it feels to die. These details reveal that memories of initial activities of the soul, just after physical death on Earth, parallel those who have had near death experiences. After numerous case studies he realised his patients were talking about a time between lives, in the ethereal or spiritual realms.

The overview of his analyses concludes that souls are generally anxious to move away from the Earth after death, but in many cases may stay a few days for their funeral. Most souls have a sense of hovering over the physical body before passing through a tunnel towards a light. They also feel a "pulling sensation" guiding them. Immediately after their death on Earth, souls are met in heaven (the spirit world) by their guides or those with whom they were close to in their previous lives. It sometimes occurs that the people they have just left are also present, although in a lower energy light emission. Souls may enter a place of restoration before they go further. They are then taken to a place where their life review or "orientation" takes place, first with a guide and later with a Council of Elders. Spirit guides continue their protective role in heaven (the spirit world) and during the orientation strongly encourage new arrivals to take responsibility for their actions on Earth. Other guides keep a regular eye on their group members while they are on Earth, and can contact or be contacted by them any time they choose. Less senior guides may even be in incarnation at the same time as their group on occasion. In fact, some will have two guides, a junior one who will themselves still be under "teacher-training", and a more senior one that appears less often and looks after far more groups.

After this review, the subjects report being transferred to a "staging area" that might be thought of as a huge railway station, in which many incoming souls are in transit to other destinations in the ethereal realms. Many report that, when leaving this area, they travel along corridors that bulge in places with "soul groups" who belong and work together in the spirit world, and who are also in transit to their own soul groups. After this, they join a group and this is where they spend most of their time in the ethereal realms. The spirit world resembles one huge school with a multitude of classrooms under the direction of teacher-souls who monitor progress. Each member of the group shares the same spirit guide, and because much of the learning process takes place through good-natured criticism between peers, the guide will be present less often as the group advances. In the course of time, all souls have the chance to become teachers and guides although, just as in the physical world, some show more aptitude and progress faster than others do.

Most of the teaching involves discussions about former lives; good and bad choices and lessons learnt and to learn. Most souls in the world today are in an early stage of development. Spirits can experience two lives on Earth at the same time and can incarnate on worlds other than Earth. A dormant part of our soul remains in heaven (the spirit realm) during incarnations. Souls learn the techniques of creation of physical items by thought, starting with simple assignments.

The subjects report that most communication in the ethereal realms takes place through telepathy, although especially private communications are effected by "touch that passes between the two souls as electrical sound impulses". Souls are able to project their former life forms to communicate with other souls. They leave heaven to start another incarnation where they enter an infant body. Ghosts are explained by Newton as discontented spirits who remain on Earth. Souls who were intent on harming others, are put into seclusion where remedial work is done on their souls. They may be incarnated as a victim in their next life.

After a time, subjects are ready for rebirth into their chosen human body. The process seems similar to that experienced at death in that they describe travelling quickly down a dark tunnel. However, there is little delay this time, and they suddenly find themselves in the womb. Apparently, the process of assimilation with the human mind of the unborn child takes some time, and must be carefully handled.

As to the timing of the incarnation, subjects report entering the child's body either only a few months into pregnancy or, in exceptional cases, leaving it to the last minute just before birth, although it seems this is not an ideal approach. Souls are not required to reincarnate, but spirit guides apply considerable pressure when the time is considered right. Souls go to a place of life selection in order to examine alternative lives to lead. Some may choose a physical disability as it accelerates spiritual evolution. There is a tendency for spirits to re-incarnate in the same geographical area they were in during past lives.

Completely novice souls appear to require about five incarnations before they join a soul group as a "beginner". They may incarnate many times over a lengthy Earth time span before advancing to intermediate status. For example, by regressing one subject right back to its earliest incarnation Newton found it had taken more than 30 000 years to reach this stage, although in another exceptional case a subject had achieved this in only a few thousand years. Group placement is determined by soul level. Secondary groups have some contact with primary groups. Opportunity for socialization and travel exists for souls living in their group. Souls are grouped with others with similar characteristics. Coloured spirit lights show different degrees of spiritual evolution. Souls have been reincarnating more frequently in recent centuries, and today souls would have had roughly two lives during the past century.

Here I must say Newton definitely loses me. If I have to die and go through what he suggests, count me as a rebel. There is simply no way I can take this seriously. I do not want to deny that there may or may not be a spiritual realm, because I think there is. My problem is that hypnosis is not a reliable method of gaining insight into spirituality. Perhaps if the information in the case studies was a little more believable it might be easier to stomach, but I cannot, in any way, accept this as credible. Newton's opinion is based on observations of subjects under hypnosis, and is not necessarily his own claims. Nevertheless, as this information comes directly from people under hypnosis as opposed to occultists, heads of religious hierarchies or some supposed spiritual gurus, these views have found favour with many who think them credible, therefore awarding Newton pioneer status of life in the spirit world.

As always, there will be those who refute such claims. Studies in cognitive science have shown that nothing about hypnosis enables a subject to transcend the fundamental nature and limitations of human memory. It does not allow someone to exhume memories that are decades old or to retrace or undo human development.

Michael R. Nash in the July 2001 issue of *Scientific American* magazine comments on hypnosis, pointing out that hypnosis is the most controversial when it comes to the issue of "*recovered*" memory. According to Nash, cognitive science has established that people are adept at discerning between actual events and mere imagined ones. However, sometimes and under some circumstances, we can come to believe (or can be led to believe) that something happened to us although in actual fact it never did.

Nash points out that one of the key cues humans appear to use in making the distinction between reality and imagination, is the experience of effort. When making the distinction between reality and imagination humans appear to be influenced by the amount of effort required. It seems that at the time of a memory being encoded it is made with a "tag", which informs or reminds us of the amount of effort we used for such memory to be formed. In the case of a memory that took a good deal of mental effort to create, we tend to interpret it as something we imagined. However, if it required comparatively little mental effort, we are rather inclined to interpret such memory as an event that actually happened to us. Hypnosis in fact creates a feeling of effortlessness and this is why people who are hypnotized are susceptible to mistakenly identifying an imagined event as something that actually happened in the distant past. Humans have the ability to change a merely imagined event into ingrained real occurrences that took place in their lives. Nash points out that there are many studies to verify this effect. "*In short, nothing about hypnosis enables a subject to transcend the fundamental nature and limitations of human memory. It does not allow someone to exhume memories that are decades old or to retrace or undo human development.*"

In summary, I would like to say that the concept of an alien soul from an alternate reality which selects an embryo for its incarnation, and selects its parents, place of birth, situation and its sex, simply does not hold water. There are far too many discrepancies in this argument. It implies that an alien soul acts as a virus or parasite and can invade an embryo that is already well developed. Even though viruses are seen to be alive, they are in the strict definition of life not regarded as being alive because of their inability to reproduce. In any case, what if a soul of some evil character such as a Hitler or some kind of lizard needs to reincarnate into the embryo of a soon to be loving mother?

What about the parents, especially the mother who surely should have a choice as to what she allows into her womb (souls apparently may fight over an embryo)? All those who claim that souls choose their parents in reincarnation also seem to believe in soul hierarchies and good- or evil-natured souls. This would mean that evil souls could reincarnate, unless the soul beings have some governing structure that will help select, but even that cannot be true since sometimes it does seem that evil souls are born.

To say a pool of souls exists in a spiritual realm and has existed since before life on Earth, waiting for life to get to a point of readiness before they started entering the bodies of Homo sapiens, is absolutely ludicrous. What if there is no soul available to enter a body, if perhaps they are all tired or taken up or simply do not like the environment and parents they are required to have? Is that person then born without a soul? After all, apparently the souls waited for billions of years before they entered the human bodies, as beforehand humans were considered unfit for inhabitation. Who can these souls be or where can they possibly come from before they managed to make this miraculous inception into the embryos? Are they souls capable of returning from "purgatory", or have they managed to do the impossible by escaping from "heaven" and "hell" (these religious concepts are discussed in more detail in later chapters). Sounds like thriller escapes from the infamous Alcatraz in San Francisco.

There are many more reasons to discredit this theory, but I will leave it at that. The concept of the soul as not being part of our own nature but rather being an alien form that invades our bodies is hard to swallow, but that is exactly what some esoteric and other researchers are claiming.

Perhaps these theories were needed to explain a specific set of phenomena. The abilities, skills and characteristics are those we would like our souls to have and the way we would like to imagine the spiritual realm. However, I find the viewpoint emanating from early occultists such as Blavatsky through to Newton's patients incredibly naïve. The groups, sessions and counselling structure Newton's patients present, sound like people who have attended too many leadership courses, business schools and counselling sessions.

Chapter 4

3 – Reptilian Anunnaki

After the previous two chapters where we examined highly advanced extraterrestrial civilisations and parasitic alien spiritual beings, we now encounter an even more supernatural, or shall I say ultra-dimensional, scenario than ever before. According to this hypothesis, human existence is not even explained in terms of divine creation or evolution through natural selection, it is generated through shrewd genetic manipulation.

Some researchers have identified creatures so highly evolved they can be both physio-biological (biological living organisms on a physical level composed of biochemical structures), as well as spiritual beings (without any physical and material form). These creatures are said to be from different vibratory levels and dimensions, although they can also be from our physical realm, with the added capacity of alternating between the two domains at will. These "alien-extraterrestrials" of three- and/or four-dimensional nature are said to have been active throughout our galaxy for millions of years with celestial skills that can make them invisible to human senses. These strange beings are still the same gods and their sons from the Sumerian mythologies and Biblical stories mentioned in the previous two chapters and referred to as the Anunnaki. Essentially, we are looking at the same mythology, but a new interpretation giving the gods extra-terrestrial physio-biological identity as well as the ability to change back to non-physical spiritual and ghost-like beings. These beings have been renamed Reptilian Anunnaki and are considered pretty evil in nature, wanting to rule and manipulate with selfish motivation. They even disguise themselves to look like earthlings, but in their actual physical form look like reptiles.

Let us look more closely at one of the main protagonists of this concept. David Icke is the author of at least ten titles including *I am me – I am free, Children of the Matrix, The Biggest Secret* and *Tales from the Time Loop.* Icke was born in England in 1952 and tours the globe addressing audiences on the Illuminati, secret societies, ancient symbolism and freedom through awareness. He combines extraterrestrials from outer space with alien spiritual beings and, with an extraordinary feat of the imagination, adds some twists and turns of his own. As explained up to this point, on the one hand we have a belief in gods who were extraterrestrials and on the other a belief that souls are from an alien spiritual dimension transmigrating and reincarnating. Icke introduces a combination creature with both characteristics.

These are not the first creatures humans have attributed with such abilities; angels have always been described as having the ability to make themselves visible for earthly visitations. Such stories are rife in the Bible and there are literarily hundreds of examples of such creatures making contact with earthlings. In Genesis 16:7, for example, an angel appeared to Hagar, a servant of Sarah, wife of Abraham, to tell her God will make her a great nation through the son she and Abraham had conceived. Sarah was already 90 years old and unable to conceive, or so she thought until the Lord visited Abraham and Jehovah Himself gave them the good news that Sarah was going to have a child (Genesis 17 & 18).

Angels also feature in modern Christianity as seen in the New Testament when the angel Gabriel appeared to Mary (future mother of Jesus) to tell her she is to conceive a child. These are only two examples of spiritual beings with the ability to make themselves visible to humans on Earth. These immaterial spirit beings must somehow possess the ability to make themselves visible to human senses by becoming physio-biological in nature appearing to be the same as ourselves.

Homo angelicansis

We see this concept is not new and does not require much convincing, as for thousands of years we have been prepared to believe such visitations are possible.

Make no mistake, some of Icke's writings are fascinating and I find his research on Freemasonry, the Illuminati and some of his conspiracy theories credible. However, his alternative and controversial views on the origins of the species are far-reaching and incredible. Sometimes, I think he even discredits his conspiracy theories with his belief in supernatural monsters. Icke believes the early root races were alien spiritual beings called Reptilians or Anunnaki from other solar systems in our galaxy who had amazing abilities due to their highly evolved status. One of his theories involves a collection of different extraterrestrials from various stars in the galaxy who intended to mine the gold on Earth and created genetically engineered humans for this purpose.

In my opinion, Icke combines the theories of alien spiritual beings (souls) from alternative vibratory levels and extraterrestrials from distant planets or solar systems forming what he calls four-dimensional aliens. These aliens can transcend dimensions by changing their physical vibration. This is quite an angle, almost requiring a belief in occultist, alien spirit beings and requiring that one consider the ancient Sumerian gods and sons of gods to be extraterrestrials rather than mere mortals. Icke's work is too lengthy to describe in detail, so I will briefly summarise the parts that are relevant to my hypothesis.

Amongst other material Icke uses the Sumerian Tablets as inspiration for his theories, but his interpretation is entirely literal. The extraterrestrials he mentions are the same Anunnaki from the Sumerian Tablets that we came across in earlier chapters, so I will not go into detail about where the name originated. The following interpretation comes from Icke's book *The Big Secret* where he also quotes extensively from the controversial Zecharia Sitchin, the African writer Credo Mutwa and the spiritual consultant and activist W.T. Samsel.

He believes the reptile species has a long connection with the Earth, going back more than 150 million years to the dinosaurs and beyond. Like lizards and snakes, they are but one form of the reptilian genetic stream in this universe. While the dinosaurs were not all cold-blooded reptiles, as modern research has shown, the reptiles and dinosaurs are closely related in physical appearance and both have spawned an amazing variety of different forms. His argument is that reptile-dinosaur genetic streams that can produce such diversity can surely manifest in a two-legged, two-armed form with a brain capacity through which a technically advanced consciousness can operate. Recently, a greater understanding of the dinosaurs shows that a hundred million years ago they were very intelligent. The bird-like reptile, Saurornithoides, for example, had a large brain and wide-set eyes allowing stereoscopic vision. It also had fingers with opposing thumbs that allowed it to catch and eat small mammals. Some studies have suggested that had the dinosaurs not been wiped out by a cataclysm about 65 million years ago, they would have evolved into a reptile humanoid by now.

Icke asks who we are to say that this evolutionary leap from classic dinosaurs to reptilian humanoids did not happen in another dimension or on another physical planet and perhaps on Earth before the dinosaurs were eliminated. Were they, in fact, eliminated? Modern palaeontology (the study of fossils) purportedly suggests that not all the dinosaurs were killed by a meteorite strike 65 million years ago and some continue to live today. He argues that although the physical part of the dinosaurs might have been destroyed, their consciousness would be retained, because consciousness is energy and although it can be transmuted into a different form it cannot be destroyed. So what happened to the dinosaur consciousness that dominated the Earth for 250 million years? Icke points out that the earliest accounts of Assyria, Babylon, Old Testament history, China, Rome, America, Africa, India and elsewhere, feature stories of dragons. What's more, the serpent symbol has also been found in ancient Britain, Greece, Malta,

Egypt, New Mexico, Peru and all over the Pacific Islands. To Icke there is also an unmistakable resemblance between some dinosaurs and ancient depictions of dragons.

By implication, Icke speculates that dinosaur consciousness survived and has become a highly evolved super and conscious spirit race with reptilian characteristics. Although he does not go as far as to say that the current Anunnaki reptilians originated from these dinosaurs, he does draw some parallels and insinuates that the two may be linked. Some 150-250 million years later, after occupying most of our galaxy or sometimes seemingly the entire universe, he believes they came back to Earth to create havoc.

In *Children of the Matrix,* he speaks about the ancient legends and accounts of "gods" from another world who interbred with humanity to create a hybrid network of blood lines. He mentions the Old Testament, for example, that speaks about the "Sons of God" who interbred with the daughters of men to create the hybrid race, the Nephilim. Icke points out that before it was translated into English, the passage used the plural, as in "the sons of the gods". He also points out that the Bible accounts are only one of many that describe the same theme. The Sumerian clay tablets, found in what we now call Iraq in the middle of the 19th century, tell a similar story. It is estimated they were buried around 2 000 BC, but the stories they tell go back long before that. In Icke's interpretation the tablets talk of a race of "gods" from another world who brought advanced knowledge to the planet and interbred with humans to create hybrid blood lines.

According to Icke the ancient accounts tell us that these hybrid bloodlines, the fusion of the genes of selected humans with those of the "gods", were put into positions of ruling royal power, especially in the ancient Near and Middle East, in advanced cultures like Sumer, Babylon and Egypt. Apparently it also happened elsewhere, according to information provided by the Southern African Zulu shaman, Credo Mutwa.

For Icke the accounts of the "serpent race" in ancient cultures are filled with serpent and reptilian symbolism in relation to the Anunnaki. He interprets the serpent from the Bible in the Garden of Eden as a story that comes from Sumerian accounts. Similarly, he claims the story of Moses in the bulrushes is a tale told about a Sumerian king long before Biblical times. These reptilian bloodlines are the origin of the "divine right of kings", the belief that only certain bloodlines have the god-given right to rule. Icke claims this is the right to rule originating from the reptilian "gods" by way of hybrid genetics. He believes these bloodlines later gave rise to the royal and aristocratic families of Europe and, thanks to the "Great" British Empire and the other European empires, they were exported to the Americas, Africa, Australia, New Zealand and right across into the Far East. There they connected with other reptilian hybrid bloodlines like those in China, where the symbolism of the dragon forms the very basis of their culture.

Part of Icke's theory is that these reptilian-human hybrid lines became the political and economic rulers of lands occupied by the European empires and continue to rule these countries to this day. The United States of America has been home to millions of people that came from an amazingly diverse genetic pool since 1776. Icke illustrates that 42 of the American presidents are related. Thirty-three of them alone can be traced back to Charlemagne, one of the most famous monarchs of France. Charlemagne just happens to be a major figure in the story of these bloodlines and their expansion out of Britain, France, Germany and elsewhere. The reptilians control planet Earth and its entire population, according to Icke.

W.T. Samsel, author of *What is the Law of One?* writes that the "gods" began to physically interbreed with the Atlantean people to introduce the royal lineage or royal bloodlines. These people were put into positions of power and ruled over the Atlantean people. According to Samsel, the royals from this bloodline were looked upon as gods by the general population of Atlantis and ruled until the first great cataclysm, which brought the Lemurian and early Atlantic age to a close. Samsel believes that the human race has been influenced and controlled for more than 70 000 years not by one but several extraterrestrial groups. Three main participants initiated this type of manipulation; one of them being the reptilians who were particularly interested in research for their own purposes.

How do these creatures do it? To be able to move around and occupy the entire universe they must have supernatural qualities and abilities, since normal physics make it impossible to travel the universe in the four dimensions of space and time as established by Einstein's physics of relativity. According to Icke, these beings are not physical but have supernatural spiritual qualities. Icke writes there are three suggested origins for the Anunnaki reptilian intervention in human affairs. The first mentioned in his book are ***Extraterrestrials***, of which he claims at least some originated in the Draco star constellation. The Draco system includes the star Thurban, the same as the North Star, by which the Egyptian pyramids are orientated. The Sumerians apparently saw the Draco star system as the dragon and called it Tiamat. The Orions are another extraterrestrial race operating on this planet, and this constellation is also connected to the reptilians, says Icke. The second origin is ***Innerterrestrials***, who live within the Earth. Icke maintains that in their physical expression, the Anunnaki are one of the many inner-Earth races who live underground in enormous catacombs, caverns and tunnels below the surface. The third is what Icke calls ***other dimensional reptilians***, that manipulate humanity from another dimension by possessing human bodies. They are from the lower fourth dimension that the reptilians control and manipulate. The reptilian control of our planet apparently involves all three of the above.

Icke believes that not all reptilians are malevolent. Many sub- and crossbreed races exist, the controlling group coming from the Draco constellation and the Anunnaki from a reptile genetic stream known as the reptilians. Icke believes the "wars between the gods" in ancient Sumerian text were conflicts and wars between different extraterrestrial races. He explains these multi-vibratory or other dimensional creatures as having the ability to transmigrate from one vibration or dimension to another like tuning into a radio or changing channels on a television set. For him creation consists of an infinite number of frequencies or dimensions of life that share the same space. Although humans are commonly tuned to the three-dimensional world or third dimension, there can be many other dimensions. Using radio as an analogy, there are many stations broadcasting but we need to tune into the right frequency to connect with different stations. Icke believes human consciousness can be tuned to other wavelengths and connected with information and consciousness operating on alternative frequencies. This is the basic principle of psychic power and it is from one of these other vibrations or dimensions that the Serpent race, the Anunnaki, is said to control the world by "possessing" certain bloodline streams. (An in-depth explanation on current scientific theories of higher dimensional realms is provided in Part 4.)

Yet, according to Icke, it is clear that the fourth dimensional reptilian controllers are themselves controlled by fifth dimensional entities. The "station" from which the reptilians operate is the lower level of the fourth dimension, the one closest to this physical world. He likens this to the lower astral dimension, the legendary home of demons and malevolent entities that go back to antiquity. His writings suggest that the non-physical reptilians were able to pour into this dimension through portals in the time-space fabric caused by the nuclear tests and explosions, which started in the deserts of New Mexico in the early 1940s, or perhaps before the cataclysms, when the world was at a more advanced technological stage than it is today.

Icke's imagination goes way off the chart, so much so that I do not think it necessary to mention all his assumptions and interpretations. However, some of his suggested influences on the human species are interesting with regard to my pursuit of the nature of our soul. Icke's ideas suggest that a creator God did not create humans nor did they naturally evolve as Darwin suggested, but are the product of a breeding programme by what seems a highly suspect bunch of spiritual and demonic creatures.

Icke, like Sitchin, also believes the Anunnaki came to the Earth to mine gold. He says they came to Africa an estimated 450 000 years ago and that the main mining centre was in today's Zimbabwe, an area the Sumerians called Ab.Zu (meaning "deep deposit"). The gold mined by the Anunnaki was shipped back to their home planet from bases in the Middle East according to Sitchin's interpretation of the Tablets. He maintains there is much more to know about the "gold mining" business, and does not believe that was the main reason they came to Earth. Although initially the gold mining was done by the Anunnaki version of their working classes, Sitchin says eventually there was a rebellion by the miners and the Anunnaki royal elite decided to create a new slave race to do the work.

According to Icke, the reptilian breeding programme appears to have produced an Anunnaki-human hybrid (Adam?) around 200-300 000 years ago. He claims the Anunnaki infiltrated Earth's population through interbreeding and so created the "instant" advanced civilisations in the low lands of Sumer, Egypt, Babylon and the Indus Valley. He thinks there was a major breeding programme in this region, probably underground, which produced a very large number of hybrid reptile-human crossbreeds. Interpreting the Tablets, he says they eventually found the right mix, which became the first Homo sapiens, a being the Sumerians called a Lu.Lu (one who has been mixed). This is probably the biblical "Adam". Lu.Lu is said to be a genetic hybrid with a function similar to a human worker bee. Icke further explains that the Sumerian name for a human was Lu, essentially meaning worker or servant or it could even describe domesticated animals. This he says is what the human race has been ever since and the Anunnaki have been ruling the planet for thousands of years, sometimes openly and sometimes in secret.

According to Icke the crossbreeding programme (via sex and perhaps test tube) is described in the Sumerian Tablets and the Old Testament (the sons of God who interbred with the daughters of men). These hybrid human-reptile lines carry the reptilian genetic code and can easily be possessed by the reptilians of the lower fourth dimension, he believes. They operate predominantly through the white races, he claims. He further mentions that there is an area of the human brain which to this day is known as the reptile brain and within this brain is the original segment to which all other parts are additions.

Well, there you go then, test tube babies the lot of us, and probably with no souls either!

My question is, if this is all true, did they genetically engineer and manipulate humans with souls, since as spiritual beings themselves they would probably know the dangers in creating slaves with spiritual qualities such as immortal souls? Or, did these extraterrestrials even have souls? Or, if these extraterrestrials are spirit beings, do they need souls since they already have the spiritual qualities of souls as described by Newton's hypnotised patients? And, for that matter, why didn't Newton's hypnotised patients meet the Anunnaki reptilians in the spirit realm during their in-between life spells? After all, both speak of exactly the same "other" and "multi"-dimensional realms but neither has met the other. Newton's patients bumped into old pals (soul mates), guides, and the likes, and Icke's creatures are evil baddies that really do manipulate and create havoc. Come on chaps, which one is it now?

This just shows how ridiculous these theories can become. Perhaps some of the worldwide atrocities can be attributed to governments, religious organizations and even some of the powerful families Icke mentions, but to claim they are the offspring of alien spiritual beings or a spiritual reptilian race is taking it a bit far.

Chapter 5

Taking the Super out of Supernatural

Houston, we have a problem! Where the heck is this going to end? Accepting the possibility that reincarnation or transmigration does exist is fine, but despite many explanations, witnesses and postulations we still do not have any account that can be considered proof. By proof I do not mean someone's individual experience, idea or philosophy; I mean justification by means of reliable methods of argument through reason and evidence. In relation to legal matters, proof is considered to be the process of establishing in a trial or legal action, through evidence and argument, the actual facts of a disputed issue. The key feature of scientific knowledge must also be supported by evidence from observation and/or experimentation. If we formulate theories, we have to devise experiments that will either confirm or refute them.

We have access to accounts and evidence submitted by people who claim to have had experiences, but the deductions drawn from these certainly are not cut and dried and usually involve stretching the imagination a little too far. In criminal cases all the facts relied upon to build the argument must be proven by the prosecution beyond reasonable doubt. A witness's evidence of what somebody else said is hearsay evidence and cannot be produced as proof in court. Evidence given by children and people with mental disorders should also be considered with caution, unless there is other evidence or corroboration to support it.

From this vantage point, we have no concrete proof that we have souls, let alone whether such soul existence may or may not be able to transmigrate between physio-biological lives. It is ludicrous for us to even consider the theory that the reptilians genetically engineered humans into our present form. Equally absurd is to think the human psyche has evolved to a point where we consider our souls to be from another realm invading us like alien parasites.

Lyall Watson, author of two best-sellers dealing with the supernatural part of human nature, writes that nothing is supernatural, it is only we that make it so. He speaks about the cluster of semi-normal phenomena that exists between matters we understand as normal occurrences, and those that are altogether paranormal that we find impossible to explain. He calls this area super-nature. Humans seem to have a need to believe in the unseen and even though this may have some merit, the question arises whether this is a part of our character or whether things unseen have real value. Watson believes that as studies about the occult increase, more will be revealed. He believes "secret knowledge" is something we have known all along but have been hiding from ourselves. His account of the supernatural is meant to extend the traditional five senses into areas where others have been operating undercover.

It becomes difficult to dismiss the unexplained when so many reports of supernatural experiences exist. Most people believe in some type of paranormal existence, whether it be ghosts, miracles, spirits, telepathy or reincarnation. Many beliefs from early civilizations are similar in nature, leading one to wonder where they come from and whether there is not some sort of common explanation. Is imagination such an inherent part of our nature that, in fact, these supernatural things do not exist but are essential to believe in?

This could apply to any form of supernatural event or experience, including the supernatural explanations in which the writers we have mentioned build their own myths from existing stories and myths. To claim we have gods as ancestors, is as supernatural as claiming our ancestors are extraterrestrials or aliens souls from other planets or vibrations, and then to pepper the theory with spiritual reptilians from alternative dimensional realms, puts the cherry on the cake. I simply cannot accept any of it. There is absolutely no proof. What does interest me, is when we take the paranormal and look for reasons to call it the soul. On the one hand I do not accept the different views that our past has supernatural beginnings, but on the other hand, if there is such a thing as the soul, then it is still steeped in mystery and supernature and has yet to be explained.

Lyall Watson believes the supernatural needs a new, fresh and thorough overview; some system designed to retrieve, catalogue and classify all unusual events everywhere. He laments the fact that many paranormal descriptions are discarded and suspects that the answers to some of the riddles of the paranormal might well lie in the pattern and content of such reports. There are no consistent findings in parapsychology, yet we feel reluctant to dismiss it entirely. Watson continues to explore the "ghosts on the edges of perception", all the while pointing out inconsistencies in natural history to find a better understanding and a more open-minded analysis of the pieces that don't quite fit.

Thus far, all our explored theories leave us with either archaic gods, extraterrestrials, alien souls, or spiritual reptilian creatures as ancestors, or else an evolutionary flow of events by completely natural selection. I go with evolution and leave the gods, supernatural creatures and spiritual beings for others. Anything more makes me guilty of either reviving or living in the old-age traditions of myths, ritual and magic. Yes, these civilizations existed, yes, they were advanced, and yes, we do not have all the answers. We have yet to find evidence to link the pieces of history together, but this is true for many other periods, and it should not be necessary to load it with supernatural mythical dogma.

Although it is abundantly clear there were earlier civilizations than the Sumerians, Babylonians and Egyptians, we know very little about them in terms of where they came from and how long they existed before our time. It is clear they existed after the Neanderthal and Cro-Magnon period of 30-40 000 years ago. What we know about the Stone Age Homo sapiens from the Cro-Magnon period is that they had a complex culture, wore fitted clothing, decorated themselves with jewellery and used instruments made from stone, bone and ivory. They were cave dwellers and appear to be the ancestors of the people of southern and western Europe during the last ice age. The mystery is how a civilization dating from 11 500 years ago managed to develop technology advanced enough to build temples during a time when the rest of humanity were still considered hunter-gatherers. According to our official version of history, it was only 6 000 years later that we developed metal tools.

Upon the findings and discoveries of ancient Mesopotamia, we base our beliefs in the soul, creation and eschatology. However, we are yet to make sense of their mythology, the cataclysms referred to and what went before, especially as current archaeological discoveries are shaking the foundations of all existing knowledge. The book of Genesis tells us Adam was the first man to exist, but anthropologists consider him the "first spiritual man". This is incongruent with earlier accounts where spirituality was recognised by people much earlier than Adam's generation. In *The Lost Testament* written by David Rohl, he mentions Adam as the original ancestor of oral tradition. He also says anthropologists would argue that the Neolithic age, which extends from about 7-2 000 BC did not produce the first spiritual being – that is to say a human who believed in existence beyond death. Archaeologists have provided evidence that late Neanderthals buried their dead with tools for use in the afterlife. Both Neanderthals and Cro-Magnons removed heads from the bodies of certain individuals to venerate the skulls of the departed within the community, or so it is believed.

But if we accept the theory that god and his sons are extraterrestrials, or our souls are parasitic alien souls, then we may as well take Genesis 6:1-4 of the Bible literally:

Homo angelicansis

"1 And it came to pass, when men began to multiply on the face of the earth, and daughters were born unto them, 2 that the sons of God saw the daughters of men that they were fair; and they took them wives of all which they chose. 3 And the LORD said. My Spirit shall not always strive with man, for that he also is flesh: yet his days shall be a hundred and twenty years. 4 There were ***giants*** *in the earth in those days; and also after that, when the sons of God came in unto the daughters of men, and they bare children to them, the same became mighty men which were of old, men of renown."* (*King James Version*)

The giants in verse 4 are also translated as "Nephilim" from the *Standard Revised Version* – Oxford University Press. The *giants* or "Nephilim", derived from the Hebrew word "nephîyl" or "nephîl", can be interpreted as *bullies* or *tyrants.* So many contradictions exist in these very scriptures and their implications. Neither is the God referred to in Genesis 6:1 the supposed Christian or Jewish "God that created the universe", nor are the sons referred to sons of such a God. God in this scripture, as translated from the original text in Hebrew, is Elohim, which is plural and should be read as gods but as shown above can be interpreted as bullies and tyrants. If we take the Catholic/Christian view that God has only one son, Jesus, then verse 4 also contradicts the core and fundamental principles upon which Christian doctrine is based. Are we now to believe that God had many sons and that Jesus was not the only one after all? Are we to believe that these sons of God had intercourse with human daughters of men and that their children are today roaming our Earth? The implication of this scripture is that Jesus had brothers and that Jesus and/or his brothers came to Earth and had sex with earthlings. The further implication is that there are now many people on Earth who can claim to be demigods and heirs to the throne of God Jehovah, since they are the direct lineage of God Himself.

If these references are taken literally and these gods and their sons were early earthling rulers, then it makes sense. But if they are extraterrestrials, it would imply that some people today are the offspring of aliens, a hybrid species of half-human and half-extraterrestrial. Some writers choose to adopt this last interpretation and as I have shown come up with some fascinating interpretations and explanations.

What utter rubbish! I can only imagine the scripture must be referring to an earthly "god" (or "gods") such as one of the tyrant kings of the time. Perhaps they were aristocracy whose sons decided to take ordinary peasant wives. No gods, angels, extraterrestrials or reptilians are responsible for our evolution, we are simply life forms evolved from Earth called humans. (See more about evolution in Part 3.)

David Icke believes an Anunnaki-human crossbreed elite was not only described by the Sumerians, but that many other accounts of the interbreeding of extraterrestrials and humans, or the "gods" and "sky people", exist. According to Zecharia Sitchin, the word Nephilim relates to "those who descended", while others say it means "the fallen ones" or "those who have fallen". Sitchin interprets Genesis 6:1-4 as referring to interbreeding between extraterrestrials or innerterrestrials with humans to produce the reptile-human hybrids. The term "sons of God", he says, in fact comes from the Hebrew "bene-ha-elohim", which really translates as "sons of the gods".

Since we do not have any written documentation dating from the Atlantis-Lemurian period (if there were such epochs), we are obliged to investigate some of the mythical mysteries from civilizations such as the Sumerians which came much later. Although Sumeria has been the official starting point of our recognised civilized history for some time, many are now refuting it due to new archaeological evidence. Their mythology talks of gods and supernatural events, as well as angels and possible extraterrestrials in the form of Anunnaki.

In describing angelic forms, the people of the Palaeolithic period (7-2 000 BC) started to recognise and document spirituality and a possible spiritual realm as an alternative reality to our own. To me this is important not because these people were convinced there were gods and angels, but because for some time preceding the Neolithic period we may have reached a stage in human development and consciousness that is attributed to gods or angels – that of multi-dimensionality.

The seemingly highly developed societies from the time of Atlantis and Lemuria may have mastered the type of consciousness that we attribute to angelic beings. However, the Homo sapiens of the time, from Cro-Magnons 40 000 years ago right up to the start of the Neolithic period were becoming increasingly aware of some kind of higher dimension, one that is superior to, or perhaps still hidden from, the one they lived in. Although not explained in scientific terms, they were instinctively and possibly psychically aware that there was more to their world than simply the sensory environment.

Some astonishing hypotheses of contemporary writers describe how the spiritual beings or souls may look. It has long been a desire for humans to give the soul or spiritual being a corporeal nature. The first documented case of the soul becoming immaterial (disembodied/ghostly) yet requiring a body as a condition of its existence was Irenaeus. He was a 2nd century Christian prelate and Father of the Church, born in Asia, who attributed a certain "corporeal character" to the soul. He represents the soul as possessing the form of its body, as water possesses the form of its containing vessel.

The way in which Newton and his hypnotised patients describe the appearance of souls contradicts the well-established views of occultists of immateriality. In *Journey of Souls* Newton informs us that a soul may appear as a mass of energy, but apparently it is also possible for non-organic soul energy to display human characteristics. According to his hypnotized subjects, souls often use their capability to project former life forms when communicating with each other. Projecting a human life form is only one of a countless number of appearances that can be acquired by souls from their basic energy core essence. Sometimes described as a rainbow, the soul is said to be recognised by the colour and depth of energy it radiates, which is reminiscent of the idea that humans have a coloured "aura". In the ethereal realms, "beginner" souls have a bright white radiance, which becomes increasingly off-white and then yellow before reaching bright gold as they progress through the "intermediate" stages. "Advanced" souls display a light-blue radiance that deepens in colour the more advanced the soul becomes.

To conclude the concept of these spiritual beings having a material form or nature I need to show the extent of the absurdity. The Anunnaki reptilians, which are also spiritual beings, as described by Icke, are usually between five and twelve feet tall. Some have tails; others do not. They have wings that are flaps of skin supported by long ribs. The wings can be folded back against the body (this is where the term "winged serpent" originates). The cape worn by Count Dracula is symbolic of these wings. The winged Draco characters as described by Icke are also known as the Dragon Race, while some of the ancient gods were described and depicted as bird-men. Some of the Draco reptilian hierarchy are apparently white or albino white and not the usual green or brown, corresponding with the description of the strange "albino" lizard humanoid in the building in Aurora, near Denver. (*The Big Secret* p 28-40.)

According to Icke, African philosopher Credo Mutwa also mentions the white reptilians. The Draco are apparently the "royalty" of the reptilians with the highest caste being the albino whites, who apparently have conical horns midway between the brow and the top of the skull. Other species, like the soldier class and scientist, are known as Reptoids. They do not have wings, but all of them are cold-blooded. The scales on their backs are much larger and they have three fingers with an opposing thumb. They have three toes with a fourth towards the side of their ankle and their claws are short and blunt. They have large, cat-like eyes that glow red and a mouth that looks like a slit. Some of their eyes are described as black and others as white with flame-coloured vertical pupils.

According to Icke's writings, the reptilians are the controllers of the so-called Greys, the classic extraterrestrial figure of modern times with big black eyes. Credo Mutwa claims to know about these creatures and adds that the "black eyes" of the Greys are actually goggles covering their real eyes, which are reptilian. That what seems to be grey skin is, in fact, an extremely tight false skin or suit. These he says, are really reptiles. I am not joking. These people really believe this and expect us to follow suit! Perhaps they are joking.

I discard the theories that view our ancestors or souls as originating from extraterrestrials and/or alien spiritual beings from alternative realms. To me it simply does not make any sense. The desire to award the soul some corporeal character has always been tempting, but I cannot help but feel that it is a sentiment as opposed to an informed opinion. The soul is either material or it is immaterial, which means exactly that – no matter or energy and therefore no colour to radiate from any such matter. In Chapter 10 I describe the occultist view of an immaterial soul and what the material part of our spiritual nature is. The soul has mostly been accepted as being immaterial and in such case cannot have any form, colour or visible energy.

PART 2

Religion, Philosophy and Occultism

Chapter 6

Before Christ – Mythology Absconded

For thousands of years humans have been fascinated by the possibility of the soul nature, what it may be and the implications of its existence. Philosophers explore the possibility that the soul may be linked or associated with the mind and its consciousness. Western religions, on the other hand, question the existence of a separate independent soul bestowed on us by God. The Catholic and Christian eschatology of the human soul maintains that the human soul is immortal and after biological death will go to purgatory until judgement day when the soul will be allocated a place in either heaven or hell.

We are not exactly sure when the study of the soul started, since our earliest records date back a mere 4-5 000 years with written records of our ancestors' earthly experience. We also do not know how the soul was perceived before early urban civilization, other than stories filled with myth and mysticism. For thousands of years some of the greatest thinkers have philosophised over what the soul is and how it influences life and death.

In philosophical suppositions, the vulnerability of a soul existence has also been highlighted for the past 2 600 years. If any of the theories presented in this discipline bear any truth, we face extreme challenges with regard to the existence and makeup of the soul. The final outcome of the principal theories precludes the presence of a soul and shows that spiritual intuition is a figment of the imagination. Any belief in the human soul existing and having any informing faculty of our thinking processes has lately been rejected. To believe in the soul as part of the human makeup is therefore indicative of a delusional mind. How then did we come to imagine we have souls and historically how did we begin defining the soul in the realm of mind science? We do have more elementary views of the soul than explained in the previous three chapters, with far less fantastical but still somewhat supernatural undertones. Let us look at how a more conservative approach to our soul nature developed.

The next few chapters give a chronological account of how philosophical reasoning and religious dogma developed before settling on their respective conclusions about the soul. Religion and philosophy developed side by side and occasionally strongly influenced one other. The two schools, however, have fundamentally different opinions on the actual makeup of human nature.

On the one hand, we have the philosophical view of mind and body known as dualism, and on the other a rather more sentimental religious trichotomous belief in the tripartite human nature of body, soul and spirit. Trichotomy characterises the soul as the third and separate part that makes up the nature of humans. This differs from the approach suggested by dualism according to which the soul is the informant of the mind, but which excludes any mention of a spirit. However, the fundamentally Roman Catholic inspired view of trichotomy does not traditionally include the mind, but lists the spirit as the third part of our makeup. Mind is conspicuously left out or assumed as categorised with either the body or soul. Trichotomy as explicitly conceived in Catholic and Christian doctrine, is highly controversial and contradictory.

The humanistic view (a phrase coined by Owen Flanagan – see Chapter 11) maintains the concept of the human soul is being challenged by the scientific model of materialism to a point at which there is simply no room for its existence. That is not to say that most people do not believe in some form of spiritual nature, but rather that our views have changed with time. For the first time in the history of conscious life, we face the possibility that the soul actually never existed in the first place. That is according to Darwin's highly respected evolutionary theory of natural selection, philosophical views of the 19th and 20th centuries, and the neurosciences of the 20th and 21st centuries. Presently, we are considered the most highly evolved animal with thoughts deriving exclusively from the brain and without a soul as a formative force or informant of the mind. What happened to the soul along the way? How did we get into the strange predicament where suddenly there is no soul? How did the concept of a soul develop in the first place?

It is significant for us to look at the events before and after Christ, particularly in view of the fact that we base our world calendar on an imaginary event (the birth of Jesus). Here we are living in the 21st century that is said to be 2 000+ years after Jesus was born and we still have no proof that this event ever happened. In any event, the A.D. dating method (Anno Domini, or "year of our Lord") was only invented in the sixth century by a monk named Dionysius Exiguus. In a later chapter I also show how the physics of space, time and alternative dimensions are still ill-defined. Notwithstanding, having a calendar determined by a fictitious birth date of an imaginary god is extraordinary.

The Breath in Judaism (and later Christianity)

It all started with a breath. In ancient times, the Egyptians and later the Hebrews (or Israelites) regarded the soul as "breath". It is fairly easy to imagine that when someone dies his or her final breath represents the soul leaving the body. This is a far cry from the modern views!

One of the greatest religious influences in the western world is that of Catholicism and Christianity. Since both use the Bible as the cornerstone of their doctrine, considering it the written Word of God, and considering that there are currently some two billion believers in Catholicism and Christianity, they have immense influence. Their belief in the soul is based on the views of early writers of the Bible thousands of years ago. These biblical views are said to be unchangeable as the Word of God cannot change and is, according to the doctrine, eternal.

The Torah and the Old Testament of the Bible are steeped in controversy and misbelief. The first five books, called the Pentateuch, were not even written by the person they are attributed to, Moses. We actually do not know who wrote these highly controversial books and yet they form an important part of what is claimed to be the Word of God, supposedly inspired by the Holy Spirit. Followers of the doctrine commonly accept that during the exodus from Egypt to the Promised Land, while on a mountain top, God gave some of this information directly to Moses. The oldest records that show how biblical writers viewed the soul nature are found in Genesis, which the Israelites and Christians would like to believe was written by Moses (around 1 000-1 500 BC). According to Moses, the first spiritual man with a living soul was Adam. We know, however, that Moses did not or could not have written the books ascribed to him. Thomas Paine, the 18th century American political philosopher, gives a brilliant account of why these books could not have been written by Moses in his book *The Age of Reason*. Not only does he show how this was impossible, but he also makes a strong case as to why the entire Bible cannot be the word of a true God.

He argues against those who say we can prove anything from the Bible (Old Testament). Paine says that before we can accept anything as proven by the Bible, the Bible itself must be proven true; for if the Bible is not true, or if its truth is doubtful, it ceases to have authority and cannot be accepted as proof of anything. Paine also comments on the many events in the Bible said to have been commanded by God that are shocking to humanity and to our conception of moral justice. In the books ascribed to Moses and Joshua we read that the Israelites came by stealth upon whole nations of people, who, as history shows, had given them no offence; that they put all those nations to the sword; that they spared neither age nor infancy; that they utterly destroyed men, women and children; that they left not a soul to breathe – expressions that are repeated over and over again in those books. Are we sure these are facts?

Paine asks whether we are sure that the Creator of man commissioned these things? The Bible tells us that those assassinations were carried out by the express command of God. To believe the Bible as the truth, we must cease our belief in the moral justice of God, for how could crying or smiling infants offend? Everything that is tender, sympathetic and benevolent in the heart of man rebels against these horrors.

In examining the authenticity of the Bible, Paine begins with the five books of Moses: Genesis, Exodus, Leviticus, Numbers and Deuteronomy. His argument is intended to show that those books are spurious, that Moses is not the author, and furthermore that they were not written in the time of Moses, nor until several hundred years afterward. He believes they are nothing other than an attempted history of the life of Moses and of the times in which he is said to have lived, and also of the times prior to his lifetime. Just as many an account is written today about things that happened or are supposed to have happened many years ago, ignorant pretenders wrote this history several hundred years after the death of Moses. The style and manner in which those books are written offer no room to believe, or even to suppose, they were written by Moses, for it is the style and manner of another person speaking of Moses.

Exodus, Leviticus and Numbers are written in the third person: "the Lord said unto Moses", "Moses said unto the Lord", "Moses said unto the people", "the people said unto Moses". This is the style and manner historians use when they speak of persons whose lives and actions they are writing about. If we grant Moses the grammatical right to speak of himself in the third person, because any man might speak of himself in that manner, we render Moses truly ridiculous. For example

Numbers, chapter 12:3 reads: "*Now the man Moses was very meek, above all the men which were upon the face of the earth.*"

In Deuteronomy, the style and manner of writing provides even more proof that Moses is not the writer. The writer then tells us that Moses was 110 years of age when he died – that his eye was not dim, nor his natural force abated; and he concludes by saying that there arose not a prophet since in Israel like unto Moses, whom, says this anonymous writer, the Lord knew face to face. Having shown, as far as grammatical evidence applies, that Moses was not the writer of those books, Paine looks at inconsistencies in Deuteronomy in terms of historical and chronological evidence. After telling us that Moses went to the top of Pisgah (and it does not appear from any account that he ever came down again), the author tells us that Moses died there in the land of Moab, and that **he** buried him in a valley in the land of Moab. As there is no antecedent to the pronoun he, there is no knowing who "he" was. If the writer meant that He (God) buried him, how should he (the writer) know it or why should we (the readers) believe him? Certainly, Moses could not tell us himself where he was buried.

Paine shows how the writer also tells us that unto this day no man knows where the sepulchre of Moses is situated, indicating that some time has passed between the death of Moses and the writing about the event. How then should he know that Moses was buried in a valley in the land of Moab? The writer lived long after the time of Moses, as is evident from his expression of "unto this day", meaning a great length of time after the death of Moses. He certainly was not at his funeral; and on the other hand it is impossible for Moses himself to say that no man knows where the sepulchre is.

In the book of Exodus, the reason given for keeping the seventh day is "because God made the heavens and the earth in six days, and rested on the seventh"; but in Deuteronomy the reason given is that it was the day the children of Israel came out of Egypt. In Deuteronomy there is no mention of the Creation.

In the 14th chapter of Genesis, the writer gives an account of Lot being taken prisoner in a battle between four kings against five. Lot was taken to Abraham, and the teller of the tale described himself as arming his entire household and marching to rescue Lot from his captors. Then it says he pursued them unto Dan (verse 14). In fact there was no such place as Dan until many years after the death of Moses. What is more, if we examine the description of the pursuit of the captors to Dan, we see it is not possible for Moses to be the writer of the book of Genesis. The place called Dan in the Bible was originally a town inhabited by Gentiles called Laish. When the tribe of Dan seized this town, they changed its name to commemorate Dan, who was the father of the tribe and the great-grandson of Abraham. To prove this, it is necessary to refer from Genesis to the 18th chapter of the book called the book of Judges.

Chronologically the taking of Laish and naming it Dan comes twenty years after the death of Joshua, who was the successor of Moses; and by the historical order as it stands in the book, it is made to be 306 years after the death of Joshua and 331 after that of Moses. Both statements exclude Moses from being the writer of Genesis because, according to both, Dan did not exist during the time of Moses and therefore the writer of Genesis must have lived after the town of Laish had been renamed. Who that person was, nobody knows. Consequently the book of Genesis is anonymous and without authority. There are many more examples of other inaccuracies, but I will not pursue the matter further.

Lastly, according to Paine, if we take away from Genesis the belief that Moses was the author (on which belief the word of God stands), there remains nothing of Genesis but an anonymous book of stories, fables and tradition, invented absurdities or downright lies. Besides, the character of Moses as described in the Bible is cruel beyond imagination. If the Bible stories are true, he was the wretch who first began and carried on wars in the name of religion, and committed the most terrible atrocities found in the history of any nation. An example is found in Numbers, chapter 31:13 when the Jewish army returned from one of their plundering and murdering excursions: "*And Moses, and Eleazar the priest, and all the princes of the congregation, went forth to meet them without the camp. 14 And Moses was wroth with the officers of the host, when the captains over thousands, and captains over hundreds, which came from the battle. 15 And Moses said unto them. Have ye saved all the women alive? 16 Behold, these caused the children of Israel, through the council of Balaam, to commit trespass against the Lord in the matter of Pe'or, and there was a plague among the congregation of the Lord. 17* ***Now, therefore, kill every male among the little ones, and kill every woman that hath known a man by lying with him; but all the women-children, that have not known a man by lying with him, keep alive for yourselves***."

Among the detestable villains who have disgraced the name of man, Paine says, it is impossible to find any greater than Moses, if this account be true. Here is an order to butcher the boys, massacre the mothers, and debauch the daughters. Paine asks what the feelings of any mother would be if one of her children is murdered, another destined to violation, and herself in the hands of an executioner. The same applies to any daughter who is destined as prey to the murderers of her mother and brother.

Paine further says that after this detestable order, the account of the plunder, and manner of dividing it, the catalogue of crimes increases: Verse 37: "*And the Lord's tribute of the sheep was six hundred and three score and fifteen; and the beeves (animals like oxen or bulls) were thirty and six thousand, of which the Lord's tribute was three score and twelve; and the asses were thirty thousand, of which the Lord's tribute was three score and one; and the persons were sixteen thousand, of which the Lord's tribute was thirty and two.*" In short, Paine says, the matters contained in this chapter, as well as in many other parts of the Bible, are too horrid for humanity to read or for decency to hear, for it appears from the 35th verse of this chapter, that the number of women-children consigned to debauchery by the order of Moses was 32 000.

Paine stresses that Moses could not have written the book of Genesis and we simply do not know who did. Since the "God of Moses" is such a controversial god, we simply cannot accept it as the work or word of a true God. Paine believes that not only the Pentateuch, but also the rest of the Bible, is suspect and not the word of a true God.

Nowhere in the books themselves is it claimed that Moses was the author. As early as the Middle Ages, Jewish scholars recognized a problem with the tradition as Deuteronomy (the last book of the Pentateuch) reports the death of Moses. The books are actually anonymous and composite works. On the basis of numerous duplications and repetitions, including two different designations of the deity, two separate accounts of creation, two intertwined stories of the flood, two versions of the Egyptian plagues, and many others, modern scholars have concluded that the writers of the Pentateuch drew upon several different sources, each from a different writer and period.

Irrespective of these contradictions, we can still say the historical belief of the people at the time was that the soul was a breath and it was only after God created man that man acquired a soul. In the Old Testament, we see that life on Earth took a new turn and for the first time the human soul was born. In Genesis 2:7 we read: "*And the Lord God formed man of the dust of the ground, and breathed in his nostrils the breath of life; and man became a living soul.*" (*Oxford – King James Version*). The soul came after material and biological life were formed and it was a complete soul. The reward for a good earthly life is heaven or the opposite of heaven; hell, a place of eternal fire and damnation with continuous pain and suffering. However, since we now know that this is not the word of a true God, we see that the people of that time recognised the soul as being part of human nature considering it to be the breath or something similar.

Strong's Exhaustive Concordance defines **soul** as **nephesh**, which is derived from **nâphash**, meaning to breathe or to be breathed upon.

Figuratively it can mean to be refreshed by a current of wind. **Nephesh** therefore means a breathing creature, like an animal, or in the abstract form it means vitality. In a wide literary or figurative sense, it could be bodily or mental. But **spirit** which is **rûwach**, means to blow as in breathe, literally to smell, or by implication perceive, accept smell. **Rûwach** therefore means wind, by resemblance breath, for instance a sensible (or even violent) exaltation. Figuratively this could extend to life, anger, insubstantiality, or by extension a region in the sky, by resemblance spirit, but only of a rational being. It can also be **neshâmâh**, which is derived from **nâsham** (to blow away like to destroy). **Neshâmâh** therefore means a puff like wind or a vital breath, divine inspiration, intellect, inspiration, **soul and spirit**.

Soul and spirit in the Old Testament are almost identical in meaning with no basic distinction between the two. However, in the New Testament, which was written much later, we see a better understanding of and an improved insight in both. This could be attributed to the more expressive nature of the Greek language and not necessarily because of a more developed insight. The New Testament was written in Greek rather than the Hebrew and Aramaic of the Old Testament.

Therefore, first there is God, then a creative act in which eventually man comes into being. This biological entity called man does not seem to have life and soul until breath (spirit) is breathed into him and he becomes a living soul with life and a personality. It is not clear whether man was already assumed biologically alive before the breath, and even then, whether he was alive before becoming a living soul. A living soul does not necessarily make the biological man alive, but in this case it seems as if it does. However, we already know that Adam was not the first biological man and therefore I will assume that the soul gave something other than biological life to Adam. I assume it gave spiritual life to Adam or a symbolic Adam and at least some of the people after this time.

Initially Judaism made no distinction between body and soul. As time passed their consciousness relating to the soul transformed and it became the principal part of our nature that could survive physical death and decay.

Mythology to Philosophy

From here onwards, we are dealing with a more intellectual attempt by renowned philosophers to resolve the mystery of the soul. However, the emphasis falls more on the nature of the soul and less on its origin and existence. To some extent I categorize the religious representatives with the traditionally accepted philosophers since both use logic and intellect to produce an argument for the soul. Unfortunately, the genuine and serious attempts by some of the creative thinkers to make sense of the soul and our origins landed in the hands of power hungry religious manipulators and politicians. Religion became orchestrated, which is really much the same thing as organised crime.

If we consider religious crimes to mean the intellectual seizure of factual elements, we can see how the institution of the church steals from humanity its ability to think. It asks its flock to believe in masterminded dogma rather than to think and use their own minds. Religion petitions its followers to believe with the heart rather than to use logic and rational thinking to ascertain truth. Faith is central to religious experience, standing in for the contradictions and unanswerable questions the believer is expected to accept. Rather than search for higher truth, believers are compelled to believe incomplete doctrine and allow themselves to be indoctrinated. Nevertheless, the thinkers and individual masterminds of spiritual (rather than religious) thought were and still are sometimes phenomenal philosophers. Their teachings were and are often deliberately misinterpreted to accommodate religious doctrines.

From little over 2 500 years ago philosophers (who were often also the scientists of the time) began to speculate about the possibility that the soul was more than mere breath; that perhaps it was a prime motivator and prerequisite for any life and without which humans simply could not have biochemical life and perhaps not even a mind. Since we are dealing with a period in which mind and logic reigned supreme, the question arose what role the soul plays in the processes of the mind. Moreover, the inquiry became a quest to find out if the soul is perhaps the mind or the prime motivator behind the mind.

The next step in the discovery of our soul nature takes us to the logicians of both the eastern and western blocks, who have many diverse views on the soul and its nature. At the height of mythos, the scientific and philosophical stage comes alive with a new kind of perception of our nature. On the one side, we have eastern and western religions with numerous influential opinions, some preferring to relentlessly cling to mythology. Then there are the philosophers who broke away from mysticism to find their inspiration from the various schools of sciences and make use of reason and logic. Lastly, there are the occultists striving for spiritual insight acquired from alternative esoteric practices. Some 2-3 000 years later, each of the three disciplines has its own unique outcome. No longer are philosophers and scientists in agreement with the mythological and religious views of the past, but a new search for the makeup of spirituality and the soul nature emerges.

The ancient Greeks from the earlier periods of the first millennia BC considered the principal component of the mind as a purely immortal spiritual essence which was autonomous and detachable from the body. This intangible concept of immortality led to the expectancy of a more tangible personal life of the soul after death. The Greeks' interest in cosmogony and theology was influenced by early Mesopotamian and Anatolian cultures. Subsequent western philosophy in turn is said to have its origins in ancient Greece, where there was much speculation about the underlying nature of the physical and spiritual worlds. Not much of the original writings by the earliest philosophers exist. Some of the viewpoints of early philosophers were documented and accredited to them by their students. There simply is no certainty about the authenticity of the authors of the earliest literary texts, but it does allow us to conduct some investigation into the beliefs of the time.

Homer is the Greek writer credited with writing the two epic poems called the Iliad and the Odyssey from the 9th century BC. Even though Homer recognises the soul and body, the soul is hardly conceived as possessing a substantial existence of its own. Nothing is known of Homer the individual, and in fact, the question of whether a single person can be said to be responsible for the creation of the two epics is highly controversial. The poems suggest that once severed from the body, the soul is much like a shadow. Even though the soul cannot sustain energetic life after death, it has the ability to survive death. The poems show that death does not constitute extinction, as the psyche survives. These early writings do not yet portray the purely non-material soul of later Greek and Christian thought, but a faded, semi-material ghost, shade, or image of the earthly man. The life of this image in the underworld is dull and destitute with an almost functionless existence.

Many of the early theorists looked at the soul from a scientific and philosophical basis rather than a religious one. Earth, water, air and fire were regarded as the primary elements of the entire universe, with a fifth element called quintessence as the informative life-giving force behind all existence. This led them to examine the elements as the substance giving life to soul. In fact, the first officially recognised western philosopher in the 6th and 7th century BC believed the elements were the motivating force behind the soul.

Thales, a philosopher from the city of Miletus on the coast of Asia Minor (now Turkey), addressed the mystery of soul in a superior intellectual manner and found it to be a motivational force. He likened soul to the element in a magnet that is able to move iron. Somehow, he conceived this substance to be water, noting how the universal act of evaporation and condensation resembled the soul. Anaximander, a philosopher, mathematician and astronomer from Miletus, was convinced it was air. Anaximander was sometimes referred to as the founder of cosmology, for his work related to the cosmos. He maintained that the first principle from which all things arise is an intangible, invisible, infinite substance that he called apeiron, "the boundless", which he also considered eternal and indestructible. Later he returned to Thales' assumption that the primary substance is something familiar and material, but he claimed it to be air rather than water. Heraclitus from Ephesus equated the soul with fire. He continued the search for a primary substance and claimed fire was the primordial source of matter, of which both the human soul and the physical world are ultimately composed. He regarded the soul as a most delicate matter, a part of the prehistoric fire.

The influence of Heraclitus around 500 BC is seen in Stoicism, which in turn had a strong influence on the early doctrines of the Roman Catholic Church. In keeping with the idea that everything is fire, Heraclitus maintained that everything in the world is in a state of continuous flux and change, even those objects that seem most solid and indestructible. For Heraclitus, the logos, which means law, word or reason, governs this ceaseless and universal change. On the basis of the logos, Heraclitus identified the laws of nature with the speech of a divine mind. His doctrine of the logos developed into the pantheistic theology of Stoicism, which is the belief that God and the material world are essentially the same since God is present in everything. It is on this kind of irrational reasoning that Catholicism and ultimately Christianity are based.

The next set of beliefs centred on an immortal and transmigrating soul. This contained the idea that the soul is immortal and able to transmigrate in and out of the worldly realm. The culmination of this doctrine is especially accentuated in the life of Jesus the Christ. According to Catholicism and Christianity Jesus was able to die, go to the depths of Hell, return to Earth in the same body three days later only to transmigrate to God's spiritual realm, and then return to Earth some time in the future heralding the second coming of Christ.

Ezekiel, the Old Testament prophet, describes the coming of the new Messiah. The essence of this dogma is that someone or something (described as a kind of spirit from another realm) will transmigrate to this realm for a purpose. To Ezekiel, even dead bones can be resurrected by the addition of a spiritual component. Ezekiel agrees with his predecessors that the soul is breath or wind, which is basically the same as air. He also concedes that allowing the soul or breath (in the form of wind) to blow over a valley of bones will give life to the bones. In principle, this indicates transmigration and immortality, since life can be given back to the dead by adding the returning breath or soul.

Most scholars date the Book of Ezekiel to the first half of the 6th century BC. His role as prophet and spiritual leader dates from about 592 BC; his knowledge of the rites of Temple worship indicates that he was a priest before the exile. Ezekiel's vision of the valley of the dry bones, one of the most famous in the Old Testament, graphically depicts how the presence of God is the crucial difference between the living and the dead.

The book of Ezekiel (37:1-14) is an indication of early pre-Christian perception of the Lord being the giver of life with the Spirit of the Lord the informative part of life itself.

1 The hand of the LORD was upon me, and carried me out in the Spirit [**rûwach**] *of the LORD, and set me down in the midst of the valley which was full of bones, 2 and caused me to pass by them round about: and, behold, there were very many in the open valley; and, lo, they were very dry. 3 And he said unto me. Son of man, can these bones live? And I answered, O Lord GOD, thou knowest. 4 Again he said unto me. Prophesy upon these bones, and say unto them, O ye dry bones, hear the word of the LORD. 5 Thus saith the Lord GOD unto these bones; Behold, I will cause breath* [**rûwach**] *to enter into you, and ye shall live, 6 and I will lay sinews upon you, and will bring up flesh upon you, and cover you with skin, and put breath* [**rûwach**] *in you, and ye shall live; and ye shall know that I am the LORD. 7 So I* *prophesied as I was commanded: and as I prophesied, there was a noise, and behold a shaking, and the bones came together, bone to his bone. 8 And when I beheld, lo, the sinews and the flesh came up upon them, and the skin covered them above: but there was no breath* [**rûwach**] *in them. 9 Then said he unto me. Prophesy unto the wind, prophesy, son of man, and say to the wind. Thus saith the Lord GOD; Come from the four winds, O breath* [**rûwach**], *and breathe upon these slain, that they may live. 10 So I prophesied as he commanded me, and the breath* [**rûwach**] *came into them, and they lived, and stood up upon their feet, an exceeding great army. 11 Then he said unto me. Son of man, these bones are the whole house of Israel: behold, they say. Our bones are dried, and our hope is lost: we are cut off for our parts. 12 Therefore prophesy and say unto them, Thus saith the Lord God; Behold, O my people, I will open your graves, and cause you to come up out of your graves, and bring you into the land of Israel. 13 And ye shall know that I am the LORD, when I have opened your graves, O my people, and brought you up out of your graves, 14 and shall put my Spirit* [**rûwach**] *in you, and ye shall live, and I shall place you in your own land: then shall ye know that I the Lord have spoken it, and performed, saith the Lord.*

Apart from being a prophet, Ezekiel must have been quite a storyteller and well versed in fantasy to try and make people believe such lies. Nevertheless, if we look at what this passage illustrates, we see that to give life to dead bones, breath [**rûwach**] and therefore spirit (Spirit of the Lord) is required. Breath and spirit are used as if they are one and the same. Not once is the word soul used. In these scriptures it seems the bones that came to life after breath or spirit was given to them do not require or acquire a soul! Either spirit and soul are considered to be the same thing, or else the soul is not required or considered as the life-giving force. As far as Ezekiel is concerned, you do not breathe soul into a body to have biological life, but rather receive spirit. The soul must then have a different function. This supports my philosophy that if there is such a thing as soul, it does not give life, but seeks life.

About the same time as Ezekiel's delusions (530 BC), the Greek philosopher and mathematician Pythagoras founded a school of philosophy. This school fused the ancient mythological view of the world with the developing interest in scientific explanation. The system of philosophy that became known as Pythagoreanism combined ethical, supernatural and mathematical beliefs into a spiritual view of life. Pythagoreans believed in immortality and transmigration of the soul, with the notion that the soul is a prisoner of the body that is released at death, and is reincarnated in a higher or lower form of life depending on the degree of virtue achieved. This underpins one of the first theories that the soul has the ability to transmigrate and reincarnate.

Pythagoras maintained that the soul goes through a series of rebirths in other bodies. His belief is that the highest purpose of human beings should be to purify their souls by cultivating intellectual virtues, refraining from sensual pleasures and practising various religious rituals. Between death and rebirth the soul rests and is purified in the Underworld. After the soul has completed this series of rebirths, it becomes so purified that it can leave the transmigration or reincarnation cycle to be eternally free. Rules were established to purify the soul (including the prohibition of eating beans and the emphasis on training of the memory). Pythagoras also believed in a "journey" of the soul, which has become a famous cliché commonly phrased by modern writers. Despite all of this philosophical rendering of soul and its abilities, we still have no clear answer of what the soul actually is.

The Eastern contemporary figure of Buddha (563-483 BC), born in what is present-day Nepal situated on the border of India and the Tibetan Plateau, is believed by his followers to be a manifestation of the universal Buddha of the cosmos. Buddha is said to be liberated from the cycle of reincarnation and empowered with superhuman insight. Buddhists believe the Buddha to be the ultimate incarnation from a series of lives. Buddhism is also unique in the history of religions, for it teaches that the individual soul does not exist but is an illusion produced by various psychological and physiological influences. Quite a view if you consider that the belief is shared by modern-day materialism. Buddhism avoids the belief in a dualistic immortal soul and denies eschatology as advocated by other religions. Buddhism is also the only religion without a God, a form of agnosticism, since it does not acknowledge or actually deny the existence of gods. If there were any, Buddhism denies them any special status or role.

According to Buddhists, humans are made up of five parts called "skandhas": the material body, feelings, perceptions, karmic dispositions and consciousness, which are all subject to continual change. None of these aggregates may be considered as permanent and are therefore unable to form or constitute a soul, which in their doctrine is called "atman". It seems strange for Buddhism to deny the soul and yet have a name and rank for it. In the doctrine of Buddha, it suggests the belief in an individual and immortal soul results in egoism that can lead to suffering. Instead, Buddha taught the doctrine of "anatman", which is the denial of a permanent and immortal soul. Buddhism specifically rejects the existence of the atman and does not believe in any conception of a soul or self that can survive death. The Buddhist view of reincarnation is altogether unclear and cannot be adjudged to be well formulated. Reincarnation is said to be simply a chain of consequences, but this subtlety is unclear and mostly lost in translation since worshippers still regard the dead as transmigrating souls. This seems contradictory as Buddhism teaches there are no actual souls to reincarnate.

The contradiction becomes more discriminating with the belief in karma. In essence, karma means the accumulation of the consequences of one's actions, good or bad, which is attached to the soul especially in the case of transmigration and reincarnation, where each new body is determined by the karma of previous lives. According to the Buddha, karma can lead to rebirth in forms other than human such as animals, ghosts, denizens of hell and the likes. Enlightenment comes through continual spiritual exercise and right living, which can shed the burden of karma and constant reincarnation. It is still only reached through an individual cycle of reincarnation, which has to reach the point at which mortal desire and suffering are surpassed.

Again we find the seemingly contradictory view of people without a soul when the soul is needed as an instructive force to preserve previous information from one incarnation to the next. Purification through reincarnation comes through opportunities to learn more of life's lessons. The main contradiction seems not to lie in the belief system itself, but rather that the philosophy is not consummated. There are still conflicting theories concerning reincarnation and the possibility of the existence of soul. The problem posed by the Buddhist view of the soul has never really been resolved and is still prevalent today.

Chapter 7

The Soul-Mind

Besides on-going investigation into the mystery of the soul, the relationship between the soul and the mind has also been examined for centuries. Does the soul inform the mind or, alternatively, can an immaterial soul be that which we call the mind?

Dualism is an important theory in the development of this subject and stands out from all previous viewpoints. In fact, for at least two millennia it was the focal point of philosophical investigation into the soul. According to this theory reality and the universe are made up of two markedly different elements that cannot be reduced or divided into smaller parts. In human nature, the mind and soul are categorized together and considered immaterial. The mind is separate from the physio-biological body, which is made up of matter. This particularly applies to humans, who are seen as two self-sufficient components: the body and the soul-mind.

In earlier times, there were no neurological studies to differentiate between the mind and the brain. The question of the mind being the product of the brain was not posed as strongly as it is today. In dualism, the philosophical question of whether mind and soul is a collective entity became a focal point of discussion, rather than the religious view held by Catholicism where a trichotomy of a differentiated body, soul and spirit is recognized.

In the 17th century French philosopher René Descartes explored dualism to its full extent in what is known as Cartesian dualism or the belief in a thinking "soul-mind", which is separate and separable from the body. The difference is that the soul-mind can exist *independently* of the body and the soul is believed to be the informant of the mind.

The earliest philosophers looked to the elements (earth, air, fire, water) as the ultimate reality. During the 5th century BC Greek philosopher Anaxagoras introduced the notion of "nous" (Greek for mind or reason). Anaxagoras believed the cause of beauty and order is the mind. He believed animals have a soul (the moving cause of things), but mind (in the sense of intelligence) appears not to exist in all animals and indeed not even in all human beings. Anaxagoras and his ideas mark a great turning point in the history of Greek philosophy, and his concept of nous was adopted by Aristotle. Although Anaxagoras seems to distinguish between soul and mind, in practice he treats them as a single entity except that he puts forward the mind as the principle of all things. What he says is that mind alone is all that is simple and pure.

At the turn of the 4th century BC Greek philosopher Democritus expounded the atomic theory of the universe as originated by his mentor, the philosopher Leucippus. According to Democritus, everything is composed of minute, invisible, indestructible particles of pure matter (atoma, indivisibles), which move about eternally in infinite empty space (kenon, "the void"). Democritus considered the creation of worlds as the natural aftermath of the never-ending whirling motion of atoms in space, where atoms collide and spin, forming larger aggregations of matter. He believed atoms make up the soul and the continuous movement of the atoms enables the entire body to produce movement.

Democritus loosely identifies soul and mind, but does not regard mind as a special faculty dealing with truth. He expressed himself more creatively than his predecessors by describing the mind as part of the soul and claiming them to be one and the same thing. This "thing" must be one of the primary and indivisible bodies. To Democritus its power of initiating movement was the result of its fineness of grain and the shape of its atoms. At this point in history, the soul and mind are united and seen as one or at least interacting as one.

Still in the 5th century BC, Heraclitus described the soul as cosmic ether or fire, the most subtle of the elements and a nourishing flame that imparts heat, life, sense and intelligence to all things in several degrees and kinds. The spherical atoms of fire are identified with soul, both having similar qualities of being able to permeate universally and set other atoms moving by their own movement. This view seems to imply that the soul is the informant of movement in humans.

Arguably the greatest philosophical personality in history was Socrates, who maintained a close philosophical dialogue with his students until his controversial death sentence in 399 BC when he accepted his fate graciously by drinking poisonous hemlock. This fourth century philosopher, also known as "The Rebel", considered soul and mind to be connected. Socrates taught that everyone has full knowledge of ultimate truth through the soul and needs only to be awakened to become conscious and aware of this knowledge. This Greek philosopher who shaped Western philosophy left no writings of his own, although Plato (one of his pupils) recorded many of his dialogues. Socrates became a major source of inspiration and ideas for generations of philosophers.

Close to the turn of the 4th century BC Plato was one of the most creative and influential thinkers in Western philosophy and although not the first, was possibly the strongest champion of dualism. The Neo-Platonist theory depicts the soul as a prisoner in a material body. Sometimes the body was expressed as the "prison", the "tomb" or even, as some later Platonists described it, the "hell" of the soul. Plato teaches the plurality of souls where the body is the vehicle of the soul, and the soul controls it like a "helmsman". The belief that the soul is distinct or perhaps freestanding from the body is evident in Plato's writings and becomes the focal point of discussion throughout the history of philosophy.

Our perception of the soul shows change throughout history as we become more conscious and new information surfaces. Plato likewise changed his mind about the soul, almost as if he struggled to formulate a proper definition of what exactly the soul is. In one of Plato's works, the *Timaeus,* there is an explanation of the origin of the soul derived from Pythagorean sources. It speaks about a world-soul, which is created according to mathematical laws and musical harmony. This world-soul is composed of two elements – on the one hand an element of "sameness" (tauton), and on the other some form of distinction or "otherness" (thateron). The individual human soul is described as being constructed likewise with a rational soul found in the head, a passionate soul in the breast and an appetitive soul in the stomach region. In the *Republic,* Plato's longest and most complex work on the nature of justice in the soul and in the state, instead of the tripartite soul just mentioned, we find the idea of three elements within the unity of the single soul.

Plato shared Pythagoras' belief that the soul is eternal, pre-existent and entirely spiritual. Once the soul enters the body, it becomes impure through association with bodily passions, but it nevertheless retains some knowledge of former existences which in essence is very similar to the Eastern doctrine of karma and transmigration. The question of immortality was a primary focus of Plato's hypotheses. The soul's assumed immateriality for Plato seemed to be confirmation of its immortality. Some of the views presented by Plato propose the pre-existence of the soul and perhaps even an eternal pre-existence.

Having established that spirituality exists, and immortality exists for the soul, Plato still claimed this was dependent on moral conduct in the present life. In some instances Plato's views had a strong theological tone in that he believed in divine judgement after death, heaven, hell and an intermediate state for penance and purification. There were even rewards and punishments and graduation was attained by one's own virtuousness. Neither Plato nor Aristotle believed humans were the only beings with souls.

When it came to transmigration of the soul, Plato and Pythagoras had similar views to those recorded in ancient Sumerian and Egyptian mythology. They thought it probable that after death transmigration of souls from one species to another or from one body to another could take place. Plato surmised that delivery of the soul from the body happened only after the soul had passed through a series of transmigrations. If the soul was of good nature over several existences, it would then be able to return to a state of pure being. If the character of the soul continually deteriorated during its transmigrations, it was believed to end in a place of eternal damnation.

Platonic doctrine introduced ideas that enabled an extreme form of transcendentalism to evolve. The conviction that the soul is an essential component in forming and maintaining life makes it impossible for it not to exist. Various lines of argument were not resolved in Plato's theories. According to him, the soul and body are separate orders of reality, and our physical existence is continuously involved in a struggle with the spiritual part of our composite nature.

Aristotle was Plato's most renowned student and, as is the case with Plato, is considered one of the leading thinkers of the Western world. He was critical of Plato's view that the soul is able to separate from the body and rather maintained that the soul (form or essence) is contained within the body it represents. He considered the soul as the form, or actuality, of the body. According to Aristotle, nature is an organic system with different species that each have a form, purpose and modes of development that define them. Aristotle wanted theoretical science to define all species with these criteria and arrange them in accordance with their complexities of form. The main degrees of form were considered the inanimate, vegetative, animal and rational. Humans were seen as having a rational soul, which was regarded as a higher form than the souls of other terrestrial species. This advanced rational soul gave humans the highest status of all living entities that die. The heavenly bodies, said to consist of an imperishable substance called "ether", and seemingly moved eternally by God, were placed even higher in the order of nature.

This concept of a hierarchical arrangement of nature was adopted by many Christian, Jewish and Muslim theologians in the Middle Ages, making it consistent with their religious beliefs. The 13th century Christian theologian Thomas Aquinas accepted Aristotle's analysis of the soul and body as two conceptually discernible elements of a single substance. In one way, this contradicts the idea of trichotomy that maintains humans have a tripartite and not a dualistic nature, but it does accept the soul as being the form of the material body.

At this point in time, the soul or psyche is officially defined as the form of a body with the potentiality for life. Aristotle's definition of the soul as "the first entelechy of a physical organized body potentially possessing life", or in other words "the life-giving force or forming principle of a physio-biological body that has the potential to give life", emphasizes the closeness of the union of soul and body. The challenge in his theory is to determine how distinct the physio-biological body is in relation to the human soul. The soul of a living being is its capability to be a part of the activities that are characteristic of its own natural makeup. Some are listed as self-nourishment, growth, decay, movement and, more importantly, perception and intellect. This describes an organism with the ability to nourish itself, to grow, to move about on its own, to perceive, think, that is alive and when death comes, decays. The capacity the organism has to execute these activities, constitutes its soul. The soul is the cause of the animate behaviour of living organisms. The soul is not separable from the physio-biological body for it is a capacity in itself and not the organism that has the capacity. According to Aristotle, there is one exception where the soul can separate from the body, and that is when it becomes pure thought that has rid itself of all personality.

Aristotle did not consider transmigration of souls between species or reincarnation of the soul possible, as he doubted the soul's ability to survive and function without a body. To him the soul is not separable from the body. This is where Cartesian Dualism (developed some 2 000 years later) differs from early dualism. The idea of a soul which cannot be separated from a body poses a problem for most religions, since it implies that unless a soul has reached the state of pure thought devoid of personality it cannot be separated from the body and must therefore die when physio-biological death occurs. The question is how the soul would survive death and transmigrate or move to the next step of purgatory, heaven or hell, as is believed by Catholics.

But for now, according to Aristotle, the soul does not exist without a body, is immaterial and not itself a body. Although not a body, it belongs to a body and exists in a body of a specific kind. Aristotle's picture differs from the Cartesian theory in that the soul is not an independently existing substance. It is linked to the body more directly and is not a separate spiritual substance encased in a separate physio-biological substance made up of matter.

Aristotle believed the spirit of an organism was the psychological principle, immortal and separable from the physical body. He somehow considered spirit to be merely "another kind of soul". He imagined a hierarchy of soul functions or activities such as the nutritive soul (plants), sensitive soul (all animals) and rational soul (human beings). In a hierarchical sense, anything that has a higher degree of soul also has all the lower degrees. All living things nourish themselves, grow and reproduce whereas animals move and perceive as well. Humans do all of the above with the added faculty of reason.

Soul has little to do with personal identity and individuality and the soul of one human is not considered any more important or different from that of any other human. One being's form is much the same as the form of any other. This led Aristotle to think that there is only soul, and not souls. People only appear to have different souls because they are different people. In some way, Aristotle's reasoning is in line with what we know about our genetic makeup, since he believed we are all different because we have different compounds of form and matter. Therefore, even if we all have slightly different bodies, we still have the same set of capabilities determined by the same soul.

Aristotle's work on the soul was phenomenal for its time. It represented one of the most comprehensive and well thought out concepts of the human soul by any philosopher. It is easy to understand how the rest of the world was influenced by his perception. Many of the other philosophies that came after Aristotle such as Cartesian Dualism, the eastern and western religions and the occult believed the soul was a stand-alone entity capable of separating from the body. In my opinion, if humans have something resembling a soul it makes sense that it should be free-standing. Even if it hasn't yet become free-standing, I would suggest it is in the process of developing the skills to separate itself from the body and brain-mind to exit the body even while the body is still alive (as seen in near-death experiences and astral travelling).

Many philosophical views that tried to accomplish a meaningful understanding of the soul and its purpose seem limp in comparison to Aristotle's works. Stoicism was a school of philosophy founded in ancient Greece. It was one of three leading movements constituting the culture and civilization of Greece called Hellenism. The stoics were influenced by many of the former philosophers and somehow managed to combine these influences with the Hebrew beliefs in "pneuma" or breath. The Stoics were way behind the great thinkers of the time, but their simplistic philosophies influenced the Romans, who became the rulers of the Western Empire, and yes, the Catholic Church that influenced world religions in the form of Christianity. Stoicism caused a split in philosophical views on the nature of soul, becoming theological and doctrinal rather than attempting to combine logic, mind and science to explain the soul.

From the 4th century BC to the rise and conceptualization of Christian philosophy after the supposed birth of Christ, four main philosophical schools existed in the western world, namely: Epicureanism, Stoicism, Scepticism and Neo-Platonism. Stoicism became the most influential school of the Graeco-Roman world, holding a more materialistic idea of nature. As mentioned previously the Stoics took the view of Heraclitus in believing the primary substance was fire and in worshipping the logos, which they identified with the energy, law, reason and providence found throughout nature. This is quite a step back from Aristotle's hypothesis. Human reason was also considered part of the divine logos, and therefore immortal. The life of reason was one of identifying with the logos. The Stoic doctrine teaches that each person is part of God and that all people form a universal family. This doctrine helped to break down domestic, social and racial barriers and to prepare the way for the spread of a universal religion.

The Stoic movement can be divided into three periods. Founded by Zeno of Citium at the turn of the 3rd century BC, he was succeeded by Cleanthes and then Chrysippus, who wrote voluminously about the subject. These three men are the leading representatives of Early Stoicism and represent the first period of Stoic philosophy. The second period, which dates back to the 2nd century BC, embraced the general proliferation of the philosophy and its introduction to the Romans. Panaetius introduced Stoicism to Rome. Among Panaetius' pupils were Posidonius of Apamea in Syria, who taught the orator Cicero. The third phase was Roman Stoicism, of which most writings have survived in quantity. Even though Stoicism had little to offer as a philosophical movement, its terminology and concept somehow became a permanent component of western religious thought.

According to Stoicism, "pneuma" or soul seems to be the aspect that gives unity and identity to a living individual. In lifeless objects, this unifying characteristic is called a "hexis" (state) and in plants it is called "physis" (nature). Even qualities of the soul such as justice are portions of "pneuma". These are classified as bodies and so have apparent causal usefulness. The Stoics hold that all reality is material, but that matter is passive and is distinguished from the life-giving or active principle called logos, which they conceive as both the divine reason and as simply a finer kind of material entity, an all-pervading breath or fire, such as the Greek philosopher Heraclitus had supposed the cosmic principle to be. These postulations are primitive compared to contemporary thinking of Aristotle's time. Living according to nature or reason, they held, is living in conformity with the divine order of the universe. They taught that all existence is material, and described the soul as a breath pervading the body. They also believed it to be divine and an element of God composed of the most refined and ethereal matter. Soul consists of the finest grained atoms in the universe, even finer than the wind and heat that they resemble, hence the exquisite fluency of the soul's movements in thought and sensation. However, the soul-atoms themselves could not function if they were not kept together by the body. If the body is destroyed, the atoms escape and life is dissolved. If it is injured, part of the soul is lost but enough is left to maintain life. Ironically, in Stoic terminology the seat of the soul is the heart and the centre of the cognitive and emotional life. This seems to be pure guess-work, as there is no evidence for any of these assumptions.

These ideas are extended with the addition of altruism. Before the rise of Christianity, Stoics recognized and advocated the brotherhood of humanity and the natural equality of all human beings as does the Catholic faith today. According to the Stoics, all people are manifestations of the one universal spirit, and all should live in brotherly love and help one another with good grace. They held that external differences such as rank and wealth are of no importance in social relationships (although preached to their congregations, this part was obviously ignored by the government of the Roman Catholic Church).

According to the paradoxical Stoic principle that all sins are equal, the sage is considered totally good and everyone else totally bad. The sage's actions resemble "proper functions" with a distinct character called "right actions". Acting purely from "right reason", the sage is distinguished by his ability to avoid passion. Normal people have morally wrong impulses or passions which are intellectual errors stemming from an inability to distinguish between good and bad, whereas the sage's evaluations are always correct. Only sages can be truly happy and free, capable of living in perfect harmony with the divine plan. Where did they get these abominable ideas?

One can clearly see the strong influence Stoicism had on Roman thought during the time preceding Paul. Paul's writings make up a huge part of Biblical scriptures, which were selectively introduced as the Word of God that became the basis for the overall view and doctrine of Catholicism and Christianity. These religions are almost a carbon copy of Stoic philosophy. The stage is set for the greatest injustice humankind has ever seen! Catholicism and Christianity! The entire religion is greatly influenced by the Roman rather than the Greek philosophical school and has hardly anything to do with Judaism where it originated. A world of fantasy which is still prevalent today.

Chapter 8

After Christ – Fictitious Religious Faith

Trichotomy is a Catholic term used to explain the tripartite division of human nature. This explanation portrays the exclusive view of Paul, who contradicts the tripartite nature according to Jesus' greatest commandment, thereby creating controversial and contentious discrepancies.

Paul was from Tarsus and was born about three years after what Christianity believes to be the birth of Jesus. He died at the age of sixty (62 AD) and is considered to be the leading missionary who almost single-handedly established the mostly western religion of Christianity. Born a Jew he was originally named Saul, after the ancient Hebrew King Saul. When the Jews dispersed into the Graeco-Roman Empire, he changed his name from Saul to the Latin Paul. Paul's letters indicate that he was well versed in Greek rhetoric and it is known that he was educated in Jewish Law.

Paul's teachings are represented in his first set of letters addressed to his own newly formed Christian supporters (followers of Paul's teachings of Jesus as the Christ) in the ancient Macedonian city of Salonica or Thessalonica at about 50 AD. They describe the perfect human (teleios) as having a tripartite nature of ***body, soul and spirit*** (soma, psyche, pneuma) (1 Thessalonians 5:23). This view, called trichotomy, is currently the official view of the Roman Catholic Church and accepted by millions of Catholics and Christians alike. However, this view is distinctly different from the presumed view of Jesus portrayed by the unknown writer/s of the Gospel of Matthew compiled some 50 years later! The first great commandment attributed to Jesus according to the scriptures as set out in Matthew 22 verse 37 is that you shall love the Lord your God with all your ***heart***, with all your ***soul***, and with your entire ***mind***. Spirit is not mentioned and the tripartite nature according to "Jesus" gives the *mind* precedence over spirit. The Gospel of Matthew therefore opposes Paul's view.

There appears to be a genuine contradiction here. It is unusual for the Roman Catholic Church to follow the doctrine of Paul rather than that of Jesus, since it professes the body, soul and spirit to be the perfect man (*teleios*), and excludes the mind. This discrepancy is far-reaching since at no time is spirit equated in any way with mind, not even in the disciplines of philosophy or the occult as I will show later. Even though this apparent discrepancy can be explained, the outcome appears to be more of a deliberate misapprehension and conspiracy than a genuine mistake.

But are we to follow the teaching of Paul or that of Jesus since, according to Catholicism, Jesus is the supposed Son of God? Jesus is even considered to be God *in the flesh* if we look at the apostle John's words: John 1:1 says that "*in the beginning was the Word, and the Word was with God and the Word was God*". Later in John 1:14 it says that "*the Word was made flesh and dwelt among us*", meaning the Word in the flesh is Jesus (the Christ and therefore the Messiah) and since the Word is also God, Word = God = Jesus.

Paul came to worship Jesus as the Messiah when, on a journey from Jerusalem to Damascus, he had a vision of what he perceived was Christ. Acts 9 tells the story of how Paul (who was at that time persecuting Jews) was confronted by Jesus and struck blind for three days during his journey to Damascus. This incident apparently changed his life so radically that he became a believer in Jesus as the Christ and long awaited Messiah of the Jewish nation.

During the time Jesus was said to have confronted Paul, Jesus would have already been dead for some time although it is reported that Paul, who was then still called Saul, and the men who journeyed with him apparently all witnessed Jesus revealing himself in the form a light and a voice from heaven. After this incident Paul was led to a disciple at Damascus called Anani'as, was filled with the Holy Spirit and regained his sight.

Other than the life-changing experience of being blind for three days, it is not altogether clear why Paul started this new theology. His complete "conversion" took three years. Paul never actually claimed to be converted (which confuses and clouds the issue), but started preaching and coercing gentiles to his way of thinking. According to Paul, accepting Jesus as the prophesied Messiah was the only path to salvation and eternal life in heaven with God.

The New Testament contains 13 letters said to be written by Paul, but at least five of these are contentious. Paul had a mysterious determination to collect money from his Gentile churches, which he would personally deliver to the Jewish Christian church in Jerusalem. His motives are not fully understood, but it appears he wished to bring the two churches together. Perhaps it was a conspiracy of some sort, but it may have been that Paul genuinely believed that he was the chosen instrument of the Messiah to bring the good news of salvation through faith to all of humanity. The problem is that he started a religion which previously did not exist, and created a person called Jesus whom he himself never met and whom we are not certain even existed. If it were not for Paul, Jesus (the Messiah) who became "God in the flesh" in all likelihood would never have surfaced. Acts describes how Paul was arrested in Jerusalem after riots incited by his Jewish opponents. After this, he was taken to Rome where he was probably executed in 62 AD.

Then follows the extremely strange part that places Christianity in serious doubt. The biographical works dedicated to Jesus according to the New Testament Gospels of Matthew, Mark, Luke and John only surfaced many decades after Paul's missionary work and his writings of Christ as the prophesied saviour. Not only did Paul preach about a person he had not met and was already dead or had ascended to heaven (making it impossible for anybody to verify his claims), but the gospels that followed decades later do the same to support the story.

This evokes the question why the Gospels appeared so long after the event, especially as they are written by unknown hands and are full of discrepancies. They contradict each other on numerous accounts and details, which render them virtually devoid of true historical value and credibility. Altogether, it casts doubt on the very existence of Jesus and his disciples. It is almost as if a conspiracy was born and a new religion formed based on the ideas and inspirations of Paul the persecutor and later adapted for the convenience of the Roman Empire. Even more suspicious is the fact that Paul's teachings were specifically directed towards the gentiles, almost as if he wanted to disperse the Jewish religious foundations. Very suspect indeed!

Paul formulated many of the ideas and terminology that were to constitute the core of Christian belief. He presented these ideas to several early Christian congregations. The four Gospels of Matthew, Mark, Luke and John, however, were written some fifty years after Paul's life and his letters to the churches. Chronologically the Bible places the four Gospels before Paul's letters to make the story more credible. However, this should be read in reverse to make more sense of the way Jesus became the newly found Messiah. This is why I place the possible life of Jesus as portrayed in the four Gospels, after the teachings of Paul. They date from almost a century after the supposed birth of Jesus.

According to Paul immortality of the soul and a place in heaven comes at a price. Romans 10: 8-10, written by Paul, teaches that you cannot achieve this goal on your own. The condition is twofold: an oral confession that Jesus is God, and faith that God raised Jesus from the dead. In other words, in order to be saved and inherit a place in the Kingdom of God (for your soul to attain immortality in heaven), you have to confess and believe as explained above. However, what you have to confess and believe is quite astonishing. Firstly, one has to believe that someone (who nobody can prove actually existed) was God. Secondly, one has to believe this figure called Jesus the Christ was raised from the dead, when the gospels corroborating these claims were written much later and are rife with contradictions.

A principal illustration of immortality and the supernatural nature of life after death is given by Paul in 1 Corinthians 15:40 onwards. The interesting thing about these scriptures though is that Paul, who was a learned scholar and known for his insight into the philosophical views of ancient philosophers such as Plato and Aristotle and the Stoics, goes against the established views by claiming that human nature is not dualistic. He maintained that human nature has a second "spiritual" life-giving component other than the soul which he said is the spirit. Nothing else about Paul's teachings is really spectacular, since it contained nothing that was not taught before. His teachings were a mixture of ancient mythology dating as far back as Sumerian tradition combined with Greek and Graeco-Roman philosophical thought from the preceding 6 centuries and, strangely enough, that of occult and Oriental nature that includes the spirit as the highest and indestructible reincarnating component. (See Chapter 10)

What is out of the ordinary is Paul's views on soul and spirit. An example of his teachings on the trichotomy of body, soul and spirit is captured in his letter to the Corinthians. In 1 Corinthians 15:42-57, Paul explains that even though Adam was initially made a living soul, it was his spirit that became the incorruptible and immortal element after physical death. For Paul the soul is added to the physical body at the beginning of one's life, only to be replaced by spirit at the end.

(42) So also is the resurrection of the dead. .... (44) it is sown a natural body, it is raised a spiritual body [**pneumatikos**]. *There is a natural body, and there is a spiritual body. 45 And so it is written, The first man Adam, was made a living soul* [**psuchē**] *as the last Adam was made a quickening spirit* [**pnĕuma**]. *46 Howbeit that was not first which is spiritual, but that which is natural; and afterward that which is spiritual. 47 The first man is of the earth, earthy: the second man is*

the Lord from heaven. 48 As is the earthy, such are they also that are earthy: and as is the heavenly, such are they also that are heavenly. 49 And as we have borne the image of the earthy, we shall also bear the image of the heavenly, so Now this I say, brethren, that flesh and blood cannot inherit the kingdom of God; neither doth corruption inherit incorruption. 51 Behold, I show you a mystery; We shall not all sleep, but we shall all be changed, 52 in a moment, in the twinkling of an eye, at the last trump: for the trumpet shall sound, and the dead shall be raised incorruptible, and we shall be changed.6 53 For this corruptible must put on incorruption, and this mortal must put on immortality. 54 So when this corruptible shall have put on incorruption, and this mortal shall have put on immortality, then shall be brought to pass the saying that is written, Death is swallowed up in victory. 55 O death, where is thy sting? Oh grave, where is thy victory? 56 The sting of death is sin; and the strength of sin is the law. 57 But thanks be to God, which giveth us the victory through our Lord Jesus Christ.

Accordingly, Paul says flesh and blood cannot go to the next destination of spirituality or "the heavenly" and refers to the **spirit** [**pnĕuma**] as the part which has that privilege rather than the **soul** [**psuchē**]. Spirit and soul as depicted by Paul in this context are influenced by the philosophies of recent Stoicism rather than that of Dualism from the preceding 600 years. He differentiates between them by depicting soul as the life-giving principle during life on Earth, but spirit as the spiritual part with the ability to go to heaven. This becomes clear when we study the meaning of the words in more detail.

Soul in the New Testament, which was written in Greek, is called **psuchē**, a word derived from **psuchō**, which means to breathe voluntarily but gently. It differs from **pnĕō**, which means to breathe hard or blow and **aēr**, which is to breathe unconsciously or to blow referring to an inanimate breeze. **Psuchē** by implication means spirit, *abstract or concrete in the animal sentient principle only*. This is distinguished on the one hand from **pnĕuma**, the word mostly used for spirit, which can also mean the *rational or immortal soul*, and **zōē** meaning life which is mere vitality, even of plants. **Pněuma** and **zōē** respectively correspond with the Hebrew **nephesh** (soul in the Old Testament) and **rûwach** (spirit in the Old Testament). (*Strong's Exhaustive Concordance*)

The spirit when mentioned in the New Testament is almost always called **pněuma**. This word derives from **pněō**, to breathe hard or blow. **Pněuma** means a current of air, like breath or breeze and by analogy it can figuratively mean a spirit, for instance (in a human) the rational soul and by implication the vital principle, mental disposition, or *even superhuman and angelic*, a daemon or divine God. (*Strong's Exhaustive Concordance*)

However, it is a thousand years after the books of the Pentateuch (I spoke at length about the Pentateuch as the first five books of the Old Testament when I described Paine's views on Moses and the slaughtering) that Paul writes to the Corinthians. One would think that Paul would at that stage know the nature of the spirit and soul, which has already been under serious scrutiny by philosophers who had made a huge impact. But by using the word **pneumatikos**, which means supernatural or even ethereal, we see that Paul re-introduces a supernatural element to spirituality and breaks away from the dualistic interpretation which includes only the soul but excludes spirit altogether. It is, however, only when Paul's letters are introduced and the New Testament is compiled that **pněuma** is used and the supernatural nature re-introduced.

The ease with which Paul's new religion is able to mix the fantastic with reality is staggering. To intertwine such implausible components of nature and super-nature by combining physical and spiritual conceptions such as Jesus, Messiah, Word, Spirit and God is fascinating. Again, the myths, ritual and magic from some 2 000 years before are recreated with this incredible emphasis on the supernatural. People seem willing to believe anything. The brain seems to be reduced to a pumping, believing heart. The mind, with logic and reason, is rejected in favour of spirit, which is based on dogma and belief systems that have no realistic value.

None of this says anything or explains much about the soul at all. To say the least, it shows complete ignorance of what the soul may be. Every time breath is used as the belief in spirit or soul, it means absolutely nothing. One may assume the view of the soul or spirit is still very primitive at this point, or else a deliberate trap is set to entice people into the new religion. Paul's views were based upon and strongly influenced by the Stoic movement that alleges the *soul* [**psuchē**] is appropriated to the purely natural life, and *spirit* [**pnĕuma**] to the supernatural life. The principle is taken further to a point where the Holy Spirit is said to be dwelling and operating in the heart. The concept of the spirit residing in the heart in opposition to the flesh, which is evil, is accentuated in Stoic belief and seen in Paul's teachings. None of this makes any sense to me.

Jesus, on the other hand, according to the Gospel of Matthew which only surfaces 100 AD, contradicts Paul as I have now shown by including the mind as the third member of the trichotomy. This view is more in line with the philosophy of Dualism that equates soul and mind. Although I mention Jesus after Paul as this is chronologically correct, we still have no proof that Jesus actually existed. The Gospels containing the life and teachings of Jesus were only written round 100 AD, which means the books of Matthew, Mark, Luke and John must cryptically appear after Paul's writings. Paul, as we know, wrote his work during the middle of the first century, long before the Gospels. If we believe the anonymous book of Matthew and if Jesus actually existed, we read in chapter 22, verse 37 that Jesus described the trichotomy to consist of heart (physical), soul (spiritual) and mind (rational). In this instance, the three are seen as separate entities. Where Jesus describes the mind as the third component, Paul maintains it is the spirit.

One of the most notable feats attributed to Jesus is his resurrection from the dead into the same body before ascending to heaven permanently or at least until the second coming. If this were true, it would be something quite spectacular and for those who saw it, proof of immortality. It is amazing how ready humanity was, during the time of Paul, for a new religion based on the supposed resurrection of a person from the dead. It is even more amazing when considering the circumstances surrounding this event: how Jesus was said to appear in the same body – one that was supposedly marred beyond recognition (Isaiah 52:14) during his crucifixion – and how he was dead for three days only to reappear intact and alive.

Some 600 years before Jesus, Pythagoras popularized the philosophy of immortality and transmigration. Plato further explored these concepts. Paul and Catholicism attempted to give credibility to immortality with the story of a prophesied Messiah (Jesus). We also have no biblical proof that Jesus actually had any intention to start a new religion. This was done later by people who had an interest in a new world religion based on the writings of Paul. Immortality of the soul through the example of Jesus has now become possible – or so Christians believe. The doctrine of Jesus as the Christ and saviour is consummated by the Roman Catholic Church. This they accomplished with help from philosophers such as Pythagoras, the Stoics and Paul.

Thomas Paine, author of *The Age of Reason*, reminds us that very little is known about Jesus historically. Jesus left no written account of himself, his birth, parentage or any other details. Nothing in the New Testament is of his writing, although the testaments are filled with information pertaining to him. The 1st century Jewish historian Josephus also wrote extensively about Jesus. The history documenting Jesus is altogether the work of other people. Historians having brought him into the world in a supernatural manner, were obviously obliged to take him out again in the same manner, as in the account of his resurrection and ascension. His miraculous conception was not something that was admitted publicity and the tellers of this part of the story had the advantage of not being identified. In any event, they could not be expected to prove the details surrounding the conception, as it is not something that admits proof. In contrast, the resurrection of a dead person from the grave and his ascension is altogether different as to the evidence it admits.

Jesus made no written contribution to the history of the time; neither did his disciples, even though they supposedly lived with him for three years. The Gospels were not written by the disciples but by unidentified writers a hundred years after the fact. They were but some of the first of many similar texts to be written over the next few centuries.

The Epistles of Paul and the Book of Acts also contain information about Jesus. Although the lack of records caused some biblical scholars to doubt his existence, today, for some unknown reason, most scholars generally accept his existence as genuine. If this is true, we consider that he may have resembled the Son of God or God Himself in the flesh and, according to the Bible, he is one and the same as God and is the only true word of God. Mixing God, associated with logos and spirit, with a biological man, associated with matter, is actually quite sick. Belief in Christianity and Catholicism requires a major leap towards "faith" rather than reason. People are asked to believe in a supernatural doctrine rather than live by knowledge and reason. This takes us right back to mythology.

With a lack of factual evidence about Jesus, we see enormous efforts to legitimize fake artefacts such as the Turin Shroud and a 1st century ossuary with inscriptions referring to James and Joseph documented as the brother and father of Jesus respectively.

The Turin Shroud is a 14 foot long linen cloth, which has remained in a cathedral in Turin for centuries and has been venerated by many faithful Christians as the actual burial cloth of Jesus. The shroud, resembling a photographic negative, shows an image of the front and back views of a man almost two meters tall, and displaying marks which can be said to correspond to the injuries Jesus suffered. Experts consider the shroud a forgery because of carbon-dating tests performed in 1988. Three independent carbon-dating tests done at Oxford University, the University of Arizona and the Federal Institute of

Technology in Zurich have shown that the shroud is definitely not the burial cloth of Jesus, since the tests date it in the mediaeval period, but more specifically between 1260 and 1390 AD.

In October 2002 there was an announcement at a Washington press conference about the discovery of a burial box with the Aramaic inscription "Ya 'Akov bar Yosef akhul diYeshua", which means "James, son of Joseph, brother of Jesus." The inscription is claimed to have been written around 70 AD. If this box or ossuary is genuine, it would be the oldest archaeological link to the possible existence of Jesus. However, sceptics doubt the authenticity of the inscription. The ancient burial casket has been declared a fake by Israeli antiquities experts. The Israel Antiquities Authority allowed several committees of experts to investigate the inscription and the box, who found that the box can be declared authentic, but not the inscription. Some scholars argue that the second part mentioning Jesus may have been added by early Christians or by a modern forger. The latter part of the inscription appears to be new and written in a modern style in an apparent different handwriting.

The owner of the box, a Mr Golan from Tel Aviv, said he bought the ossuary in the 1970's from an antiques dealer in Jerusalem for about $200. Golan was arrested and questioned about the ossuary by the Israel Antiquities Authority. He denies that it is a fake, or if it is, that he had anything to do with forgery. An ossuary like this one, especially with such significance, increases the value of an artefact and while Golan may have only paid some $200 for it, the ossuary was insured for $1 million on a trip to Toronto during which, incidentally, the artefact was severely damaged. Even though Golan says it is not for sale, estimates show that its value may be as much as $2 million.

Nevertheless, that is not the end of the story. There are other ossuaries with perhaps more meaningful inscriptions, but they do not make the headlines and are hardly mentioned. One in particular is said to have "Jesus, son of Joseph" written on it. This finding was only published a few times and never came to the attention of the academic world at large. I imagine the reason for this is that it places the entire existence of Christianity at risk. If it were true it could mean the end of the supernatural ascension of Jesus as is believed by Christianity today.

The validity of a true artefact of this nature would have enough power to destroy the creed of Christianity altogether. The fact that there are actually two such ossuaries may have various implications. It could mean that the names of Jesus and Joseph were so common that there were other families with such relations. It may also be that both or at least one of them is fake, which puts us in a similar situation as the James ossuary.

According to Paul Flesher, director of the Religious Studies Programme at the University of Wyoming, two ossuaries have Aramaic inscriptions reading "Jesus, son of Joseph". Flesher has been researching the James ossuary inscription and while reading the entries in a catalogue of nearly 1 000 ancient ossuaries, he came across two with the inscription relating to the exact name of Jesus as it would have been written at that time. One discovery was made in 1926 and the other was a family tomb unearthed in 1980 during excavations at East Talpiot in Jerusalem, where several ossuaries form part of a collection that seemed to be from one family. The London *Sunday Times* published this discovery in 1996, 16 years later. Archaeologists then moved the tomb's ossuaries into museum storage at Romemma. The ossuaries were individually inscribed with the names Jesus, son of Joseph, Mary, Joseph, Jude (a possible brother of Jesus), and Mary (possibly a sister of Jesus). Nobody who deciphered the inscription or those who catalogued the boxes seems to have thought the inscription could be referring to Jesus Christ.

All the ossuaries were devoid of bones since Jewish laws are very strict in Israel and artefacts containing bones are sensitively reburied by the authorities, although their containers may be removed. Outside the archaeological context that might prove or disprove the inscription, these ossuaries are devoid of any real meaning and have similar value as the Shroud of Turin since we are unable to prove their authenticity. The fact therefore still remains that we cannot verify the life and death of Jesus the Christ.

The Roman Catholic Church is the largest single Christian body with Christian believers who acknowledge the supreme authority of the Bishop of Rome, the pope. This Church regards itself as the only legitimate custodian of the Christian creed by an unbroken succession of bishops descending from the apostle Peter to the present time. It considers itself heir to the theological supposition of the apostolic, patristic, medieval and modern periods, starting right from the earliest Christian communities. Theologian Thomas Aquinas taught that “scripture alone” is the source of theology and for Christians, the Bible is accepted as the basis for its teaching.

It is impossible to separate the history of the Roman Catholic Church from the history of Christianity, even though a break with the Eastern Church in 1054 AD and another with the Protestant Churches in the 16th century changed the face of Christianity. The characteristically Christian figure of the bishop had emerged by the middle of the second century. The recognition of Christianity by Emperor Constantine in 313 AD consolidated the Church. In the Christian eschatology the second coming of Christ, the resurrection of the dead, the last judgment, immortality of the spirit (not the soul), heaven and hell, and the consummation of the kingdom of God are all part of an astonishing and incredible doctrine created over many centuries. The creed of Christianity also supports the belief in the Son of God who will judge the living and the dead when he returns, and a reign of a thousand years on Earth before the end of our existence as we know it, accompanied by the resurrection of the just and the unjust to stand trial at the last judgement day.

These fantastical stories are exactly what we need to get away from. There is no truth in them, they have no merit and should die a quick death right here. I do not need to go into great detail or write volumes about the reason, for this has already been done by many accomplished scholars. Suffice to say it is not the truth, never was and never will be. Soul or the life and existence of soul have nothing to do with religion. The doctrine described previously is a myth or part of mythology as much as the ones created 3-4 000 years before the so-called birth and life of Jesus. It is nothing more.

Most Christians believe that each individual has an immortal soul. They believe the human personality as a whole (composed of soul sustained in purgatory and a newly resurrected body), may through faith be granted God's presence in the afterlife after the final judgement day. The idea of transmigration was never successfully adopted into Christianity. The Gnostics and the Manichaeans did initially believe in transmigration, but early Christians who adopted Gnostic and Manichaean doctrines were declared heretics by the Church. I find this strange though, since the resurrection and second coming of Christ in effect means exactly that, transmigration and reincarnation in and out of the spirit realm and back. The same goes for the second physical body that all successful and saved souls will receive at judgement day before going to heaven. It all relates to the same concept of reincarnating from one realm to another and in and out of one physical body to another.

Genesis says that God breathed life into man, and so he became a living soul. Man became soul and not spirit. It also indicates that this process may have applied to humans only and not to any other species. We may be the only species conscious of our own souls and able to manipulate and do things in favour of such a soul. Paul, in the New Testament, adds an immortal spirit to that soul but seemingly only at the end of life.

According to Genesis 2:7, man was created first and the introduction of the soul came thereafter. Perhaps, because the soul came after man, the common belief is that the soul came from an alternative reality. The breath of life which now seems to have resulted in the soul was added directly afterwards, almost as if the corpse was lying there as a lifeless body, but once the soul was added the body of Adam became alive for the first time. The problem with this scenario especially in Christianity is that man was created in one day, the sixth, and that it must have been at some point on the same day that the soul was breathed into man's body to give it life. This contradicts our scientific knowledge of human evolution and cannot actually be true. It does, however, significantly resemble the point in evolution where humans become aware of their new acquisition, the soul.

Origen, the celebrated 3rd century biblical scholar and writer of the early Christian Church, taught the principle of trichotomy, corresponding to the threefold division of the person into body, soul and spirit, which by then had become a common concept. He is credited with developing the idea of Christ as the Logos, or Incarnate Word, who is with the Father from eternity, but he also taught that the Son is subordinate to the Father (God). He adopted the philosophy of the pre-existence of the soul, which was severely criticized by many of his contemporaries and by subsequent writers. Theories that were developed from his doctrines became the subject of considerable theological controversy during the Middle Ages and even today are not accepted by many who follow the creed of Christianity. Origen's teachings of the pre-existence of the soul are similar to Hinduism, occult teachings and current beliefs in Alien Souls that pre-exist before coming to Earth.

Plotinus, also from the third century, was a Roman philosopher and the founder of Neo-Platonism. Plotinus accepted a doctrine of emanation and believed the soul is part of the third and lowest member of the Trinity – the Holy Spirit. Plotinus believed the soul creates the material world from memory of the divine, with something analogous to sexual desire as motive rather than reason. Matter is created by the soul, and cannot exist independently. The soul is neither the matter nor the form of the material body; it is rather the eternal and immortal essence. His doctrine speculates that there is a constant transmittance of powers from the Absolute Being to the creation through several agencies such as "nous" or pure intelligence, from which the soul of the world flows. From this, in turn, the souls of humans and animals, and finally matter, flow.

According to Plotinus the soul has a dualistic nature with the inner soul intent on "nous", and an external manifestation in which the soul generates its image. This seems to imply that the information from the senses is linked to something other than the soul. However, he considers sensory input different to pure intelligence, which to him is intuition of the nous, something acquired through intellectual meditation and not merely impulse from the senses. I interpret this as enlightenment or even trichotomy where the soul is considered to be above the realm of the brain function, which is a material object. This is quite a step towards our perceptions of the mind being the product of the brain, which is a view based on modern neurobiology. Plotinus' doctrine also entails reincarnation, since the soul descends and enters the body several times. The sin of the soul must be punished by entering another body. If you have murdered your mother in one life, you will in the next life be a woman and be murdered by your son.

Constantine the Great, who lived from 274 to 337 AD, was a Roman emperor and the first Roman ruler said to be converted to Christianity. He had previously been a pagan solar worshipper, and once converted looked upon the Christian deity as a bringer of victory. Constantine intervened in ecclesiastical affairs to achieve unity in the Roman Empire by presiding over the first ecumenical council of the Church at Nicaea in the year 325 AD. The previous three centuries after the teachings of Paul and the rise of Christianity through the writings of the Graeco-Romans had set the stage for Constantine to formalize the last obstacle of the Roman Catholic Church.

This rogue forcefully promoted Jesus as one of the tripartite deities of the current trilogy of Father, Son and Holy Spirit. It was accepted and incorporated into the Catholic and Christian doctrines and has been considered as Divine Truth ever since. By the time the Council of Nicaea took place Constantine was said to be completely Christian, but still tolerated paganism among his subjects. It was a Roman emperor and sun worshipper that forcefully prescribed at the Council of Nicaea that all denominations in the Roman Empire must consolidate their doctrine and teachings by proclaiming that Jesus Christ is God, the foundation and possibly the cornerstone of all of Catholicism and Christianity. Without this implicit instruction by Constantine, the figure of Jesus would today still have been considered a mere man.

The next principal architect of Western theology was Augustine of Hippo who lived at the turn of the fourth century AD, and whose literary efforts did more than any other body of writings, except for the Bible itself, to shape the Christian religious system. After the above-mentioned abomination, the Bible in all likelihood had to be re-written to accommodate the lies and contradictions. Augustine was one of the main Catholic perpetrators who formulated the doctrine of original sin that has heavily influenced the entire Christian movement. As a religious pawn of the Roman Catholic Imperial Church, he seemed to be as uninformed about the soul as were the rest of the religious world. But then I don't think he was supposed to know much about the make-up of the soul.

Augustine was born in what is now Souk-Ahras, Algeria and was allegedly one of the greatest of the Latin Fathers and one of the most instrumental of the Church. He brought the systematic method of philosophy to Christian theology. It is difficult to understand how he succeeded in doing this when the basis from which he worked was wrought with discrepancies and lies. The Roman Catholic Church has found special satisfaction in the institutional or priestly aspects of the doctrines of Augustine. Roman Catholic and Protestant theology are largely based on his strict theological opinions. He held that human nature is in a state of sin due to Adams disobedience and this renders humans powerless to change. In his theology, we are saved only by the gift of divine grace. He strongly defended the place of free will in collaboration with grace. Later in chapters 9 and 11 we will see how free will is rejected by current scientific and philosophical views.

Like Aristotle, he makes the soul the final cause of the body. As God is the good of the soul, so the soul is the good of the body. Augustine also contributed a good deal to spiritual philosophy by utilising the doctrine of emanation. God the "primeval" or "eternal One" brings forth by emanation something called "nous", which is seen as intelligence or the faculty of reason. From this intelligence in turn springs psyche or mind, which is also soul. This is an image of "nous", but also distinct from it. Matter is a later emanation. Soul has relationships with both ends of such reality and its perfection is achieved in turning towards the "Divine Trinity" from which it came. This is the same Divine Trinity which was fabricated by emperor Constantine at the Council of Nicaea in 325 AD. Thus Augustine's entire doctrine is based on untruths, extended into a philosophy.

The belief in a human soul is central to several philosophical and religious systems in eastern religions. Muhammad, the chief prophet of Islam who lived at the turn of the sixth century AD, is sometimes described as the initiator of Islam. In Muslim religious context, the pure monotheism of Allah (God) has been revealed through many prophets from creation before and up to Muhammad. Islam adopted from Judaism and Christianity the doctrine of a coming judgment, a resurrection of the dead and everlasting punishments and rewards.

The teachings of Islam on the soul resemble those of Judaism and Christianity. According to the Islamic holy Koran, God breathed the soul into the first human beings and at death, the souls of the faithful are brought near to God. In addition to the physical world, God also created angels and spirits. The angels have various roles such as the delivery of God's revelation to the prophets. The spirits are usually known as the "jinni" that inhabit this world and they can affect human beings in several ways.

The angels neither eat nor drink and are believed to be free from all sin. There is no distinction of sex among them and as a rule they are invisible except to animals, although sometimes they can appear in human form. The principal angel is Gabriel who is seen as the guardian and communicator of God's revelation to man. Michael is the guardian of men while Azrail, the angel of death, has the duty to receive men's souls when they die. Israfel is the angel of the Resurrection. It is believed all men will have to rise from the dead and surrender to the universal judgment. The Resurrection will be general and will extend to all creatures from humans to angels. After death, each human being will be judged and will either receive salvation or be irrevocably condemned to eternal damnation. I have simply given a summary of Islam since it resembles Christianity to a large extent, but is more extreme in certain respects.

Neo-Platonism, advocating the soul as prisoner in a material body, prevailed in Christian thought until the 13th century theologian Thomas Aquinas accepted Aristotle's analysis of the soul and body as two conceptually discernible elements of a single substance. Aquinas' works have made him arguably the most important figure in scholastic philosophy and one of the leading Roman Catholic theologians. However, there seems to be confusion in his acceptance of Aristotle's dualistic approach if seen in the light of the Roman Catholic doctrine of trichotomy and the tripartite human nature of body, soul and spirit. At this point, soul is debated at length and in lofty philosophical language. One tends to think the subject matter became merely a tool for mental gymnastics rather than an attempt to clearly understand the makeup of the soul and its genuine existence. The soul, its makeup and relation to life is taken out of the realm of spirit and put in the realm of the mind.

According to Aquinas the rational soul has a sensitive and biological component or body and is seen as the form (and possibly informant) of such body. The soul is an incomplete substance in that it must exist in the physical or vegetative body and makes up the substantial unity of human nature. Although connected to the body, the soul is still seen to be of spiritual nature. It is not actually immersed in matter and its higher operations are, as such, independent of the material or vegetative being. But the rational soul is supposedly produced by some special creation at the moment when the human being is advanced enough to receive it. In the first stage, when the embryo is developed, it has merely vegetative powers, after which a sensitive or sensory soul comes into being and later yet this is replaced by a rational soul, which is essentially immaterial and is implied to be the result of a special creative act.

The strange thing is that somehow these Roman Catholic gurus must have forgotten that the soul according to their Bible is "breath" and "breath" alone. Since the Bible is their only true "Word of God", they should not be tampering with it and turning it into something else. The soul was merely added to Adam in the form of a breath and anything else added to this, outside the context of the Bible, is supposition and hypothesis. But, as we can see, there are many contradictions in religion.

Chapter 9

Our Incomprehensible Soul

The concept of dualism was a prominent feature of philosophy up to the seventeenth century. René Descartes, the French philosopher, scientist, and mathematician often referred to as the founder of modern philosophy, explored it to its fullest extent with Cartesian dualism. Where dualism regards the soul-mind as attached to the body and incapable of being separated, Cartesian dualism sees the soul as free-standing and separable.

Cartesian dualism introduces the idea that the mind is part of, or perhaps even intrinsically the same as, the soul. The trichotomy theory (body, soul and spirit) is rejected in favour of a dualistic view that only recognises body and mind. According to this view the mind and body also contains the soul. If two components such as the body and the mind have the ability to be separate (much like having thoughts even without a body), then it follows that the mental and physical models must be distinct from each other. Even though this soul-mind can direct matter, it is still separate from the body it directs. Descartes does not have an answer to how the soul-mind is capable of directing matter. In other words, how an immaterial element is able to act on and direct matter to execute commands.

During the time of Descartes, philosophers started asking more probing questions about the nature of the mind/body connection. If they are separate entities, at what point do they connect and interact? There are many theories and –isms that deal with the questions posed by Cartesian dualism, but none of them clearly define how the mind/body connection actually works. For instance, how does the mind know about the body, can it influence the body and how does the body affect the mind?

Famous for his statement of "Cogito, ergo sum", "I think, therefore I am", Descartes' theory still shapes the development of psychology. For Descartes, self-awareness is sufficient evidence to imagine the mind as independent of the body. He strongly believed in the existence of God, whom he claimed created two kinds of substances that make up our entire reality. He conceived the soul as a thinking, conscious substance located in the pineal gland and the body as an extended substance with an essential characteristic extending over a certain amount of physical space. While extended substances act in accordance with the mechanical laws of physics, thinking substances act in accordance with the laws of thinking.

He maintained that our minds are not our brains. Our minds exist without the need for spatial location and continue to exist after the death and destruction of our bodies. But then, what is it that connects the mind and brain? How can a non-physical mind cause a body to move? How can information about the external world be received by our senses and cross the line that divides the physical and the mental enabling us to consciously perceive things outside ourselves like other living beings and objects? His answer is that the state of our minds causally interacts with the state of our brains. When we have aches and pains that cause us to moan or contort our faces, it is caused by a brain state, which in turn causes bodily movements. Through deliberate action we act on our desires and intentions, which cause brain states that cause our bodies to move thereby influencing the physical world. The physical world then influences our minds through its influence on our brains.

Perceptions of physical senses were said to involve causal transactions from the physical to the mental. Descartes stated that a two-way psycho-physical causal interaction is evident. When this takes place from the mental to the physical, it is an action, and from the physical to the mental a perception. The state of being joined in Cartesian dualism is not resolved, but superseded by the doctrine of two-way psychophysical causal interaction, which is called Cartesian interactionism – the theory that speculates the mind and body are capable of acting on or influencing each other.

Cartesian dualism also makes the assumption that we have a free will which is determined by something akin to the soul. Interestingly, the supposition of free will is a very contentious point of debate amongst philosophers. Those opposing this view see the mind as physical in nature and the product of the brain. Their argument is that since all subjective thought is based on objective impulses from the outside, there cannot be such a thing as free will. At this point, nobody seems able to convincingly argue the state of being or the connection between soul/mind and brain, forcing some philosophers to revert back to the old tales of mythos and faith.

Many scholars found Descartes' theory difficult to accept, claiming that mind and matter are inherently incapable of affecting each other and that any mutual action between the two is caused by God, who, on the occasion of a physical change, produces a corresponding mental change or vice versa. Other followers of Descartes abandoned dualism in favour of monism, which maintains that ultimate reality is composed entirely of one substance which for most is matter.

After battling to solve the difficulties presented by Descartes' theory of mental causation, two 17th century philosophers, Gottfried Leibniz and Nicolas de Malebranche, settled on a view that involved preserving mind-body dualism and abandoning Descartes' conception of free will. Both Leibniz and Malebranche tried to separate the thesis of dualism which claims that humans partake of two realms (one mental and one physical), with the thesis that these two realms interact. To them they were two different studies with different outcomes.

Malebranche went on to develop a philosophy called occasionalism, which denies the possibility that matter can in any way act upon or interact with mind. His was the first important attempt to deal with

Descartes' contradiction. Malebranche argued that since human knowledge is possible only through interaction between a human being and God, mind and body do not interact but God makes them appear to. Occasionalists identify force with the will of God. In its most extreme version, occasionalism is the doctrine that all mortal created entities are devoid of fundamental worth or value, and that God is the only true connecting agent. Bodies cannot cause effects in other bodies or in minds and minds can in turn not cause effects in bodies of any kind, or even within themselves. God is directly and solely responsible for bringing about all phenomena. He held that both mind and body are causally ineffective and that God is the one and only true cause.

Leibniz maintained that God set out a plan for each person's mind and body to run in parallel. His "pre-established harmony" theory refuses to admit any inter-causal relation between mind and body as suggested by Cartesian internationalism. Sometimes known as psychophysical parallelism, it teaches that God has a master plan with the non-physical nature and physical nature running parallel to each other. Parallelism simply accepts that every mental event is correlated with a physical event in such a way that when one occurs, so does the other. In parallelism, the mind and body exist in a harmony that has been pre-established by God from the moment of creation.

Another 17th century philosopher, Baruch Spinoza, taught that both material and spiritual phenomena are attributes of one underlying and infinite substance. According to him, soul and body is one and the same thing, a form of expression of "divine essence". He argued that since substance is completely self-sufficient, and only God is self-sufficient, God is the only substance. In other words, God is everything. That would mean God is the universal essence or nature of everything that exists. His view is very similar to Monism, which considers ultimate reality to be composed entirely of one substance. In order to retain the notion of God as the one true cause without sacrificing the idea of causes and effects as functioning in both the mental and the physical domains, Spinoza abandoned Descartes' view of two components in favour of what is called the double-aspect theory. Double-aspect theories are based on the notion that the mental and the physical are simply different aspects of one and the same substance.

Death of the Soul

The quest to merge soul with mind and establish soul as the informing agent for mind becomes seemingly impossible. In addition, to establish the point of association between mind with matter, or fuse soul-mind with matter, leads to a philosophical dead end. In essence, the philosophical meaning of soul is removed from the study of mind science. From here on, the all-physical reality (even if in the guise of material form) becomes a component of the mind. The assumed existence and role of soul as the informant of mind is rejected altogether.

Without having established if a non-material phenomena can interact with a material body and cause it to respond via command or will (which is seemingly created in the mind), then more knowledge of the nature of mind and body or soul-mind and body is required. Without further knowledge, the discussion will lead us nowhere.

George Berkley was an early 18th century immaterialist who believed all material things are collections of mental ideas. He denies the possibility of material substance or physical reality existing without first being perceived in the mind. From this perspective, mind/body distinction cannot exist because body is merely the perception of mind.

Another 18th century German philosopher, Immanuel Kant, who was highly influential, believed that the soul was only the subject of the processes of the consciousness and not a separate entity. What this means, is that the brain is accustomed to summing up phenomena as terms. It eventually perceives these terms as reality, but they are only the result of its own activity.

Homo angelicansis

The concept of dualism fails us in that it does not offer a logical framework indicating that the soul can exist, whether it is seen to be the mind itself or merely the informant of the mind. Nor can we say with certainty that the soul, if it does exist in some form or other, can stand separately from the body or even direct the body. After 3 000 years of rigorous religious and philosophical argument about the existence of the soul, its makeup and function has yet to be determined and defined. It all boils down to a matter of faith, and this I cannot accept.

The search for the soul continues.

Chapter 10

Occultism Revived

Religions such as Judaism and Christianity suggest that before Adam received the breath of life and became a living soul, he was first a biological man. These doctrines therefore teach that the human soul was a belated addition to the biological body that was formed from the dust of the Earth. However, the soul will not "return" to a spiritual realm such as heaven or hell, since it did not originate from such a realm to start with. Esoteric and occult teachings differ from this view by saying the soul did in fact first exist in a spiritual realm and that the material body formed as a consequence of the soul's desire to have some form of physical experience. Occultism, as most eastern religions, regards some form of spiritual component as the initiator of life, which is able to transmigrate and reincarnate to the physical realm over many lifetimes. This is as opposed to our souls having to transmigrate to the spiritual realm. The two viewpoints, in essence, are a complete antithesis of each other and obviously hugely contentious.

In the scientific arena, the possibility of our soul nature is seldom acknowledged and has never been included in any specific model or experiment to establish its authenticity. The concept of eschatology and the soul's destiny in the form of purgatory, heaven or hell can in no way be verified and therefore has no significance. Yes, it feels good to speculate about and postulate the soul's existence, but it certainly does not hold water in terms of a scientific model or philosophical reasoning. In eschatology, the soul's credibility serves the purpose of enforcing an impression that enables religious organisations to captivate and indoctrinate the imaginations of people. However, religious doctrine holds no water since it is riddled with dogma, which is sustained merely by faith.

Western occultism has much of its roots in ancient Mesopotamia and Egypt, but is also strongly influenced by eastern religion from the contemporary Indus Valley civilizations, which flourished from 4 600 to 3 800 years ago. Nearly all traditional civilizations have occult teachings with transmigration and reincarnation of a spiritual component at the heart of their doctrines. The 19th century saw a huge revival of occultism with the advent of the Theosophical Society and the Hermetic Order of the Golden Dawn, with rebirths in the 1960's and the start of the New Age movement in the 1980's.

The term occult refers to the study of hidden or invisible things, which are of esoteric nature, but the term is sometimes misinterpreted as knowledge or phenomena hidden from man by God or Satan. This makes the word and its association somewhat sinister with many people avoiding it out of fear or ignorance due the impression that it has dark connotations. The real meaning is not dark at all and merely suggests that previously hidden things can be made obvious through insight and enlightenment. Esoteric observations are mostly spiritual in nature and enlightenment on such matters mostly individualistic. Not everybody seems capable of making such observations and hence the mystery surrounding the study of occultism.

About the same time that our philosophers discard the soul, occultism makes a new and fresh stance in its favour. Just when you think the philosophers have led us to a spiritual cul-de-sac by not uniting mind and soul, Helena Petrovna Blavatsky and Rudolph Steiner come on the scene with a strong stand in favour of a soul existence rooted in occult belief. Leaning heavily on eastern beliefs and on challenging scripts like the Sanskrit and Vedas, Blavatsky in *Isis Unveiled* and *The Secret Doctrine* tries to regain spiritual knowledge for the West. By using far more familiar layman's language, Steiner manages to make esoteric knowledge more comprehensible and accessible to his students in various lectures and books on occult science.

Occultism relies heavily on the ability to access information of a paranormal nature outside the range of the physical senses. This I refer to as being "clarient" or having "clarience". These terms cover clairvoyance, clairaudience and clairsentience in one word – by which we are able to access paranormal (seemingly spiritual) information without the use of the senses or outside the range of the physical senses. This is ostensibly impossible from a scientific point of view since it cannot be supported by any substantial evidence.

Clarience can be compared with extrasensory perception (ESP) or precognition, which includes any ability to gain information by psychic means rather than through the physical senses. However, in the case of ESP and precognition, access to information can include telepathy from one person to another or communication with guides or spirits. I do not regard the existence of guides and spirits as genuine and therefore do not accept this information as credible. ESP and precognition also covers information about future events, which again I do not include in my definition. I use the term clarient or clarience simply as an independent and self-gaining ability to access information outside the scope of the physical senses and seemingly without the help of spirits or supernatural beings.

The word soul is often confused with spirit, but in the occult sciences the distinction of soul clearly infers that it serves as a mediator between spirit and the material body. Similar to Christian trichotomy, humans are seen to have a triune nature with a physical body and eternal spirit connected for the duration of a lifetime by the soul, although the tripartite nature is somewhat more intricately defined. In the *Oxford Dictionary* spirit is defined as "*the animating or life-giving principle in a person or animal*", "*the intelligent non-physical part of a person; the soul*". This according to occult belief is not accurate as we will see later in this chapter. It is also amazing that the *Cambridge Dictionary of Philosophy* says the same when referring to "*soul, also called spirit, an entity supposed to be present only in living things, corresponding to the Greek psyche and Latin anima*". The synonymous use of soul and spirit is partly due to the similarity of the use of soul and spirit in ancient texts such as the Bible.

Human Nature

In Theosophy (which is a study of occult nature), there are seven "principles" or planes that suggest the seven individual and fundamental aspects in a human. In ascending order, they are the physical or biological body, the Etheric body, Astral body, Ego or "I", Manas, Buddhi and Atman. These are divided into two ternaries, the higher and lower. The lower ternary consists of the physical or biological body on the physical plane, with the Etheric and Astral bodies that cannot yet commonly be experienced by the five senses. The Manas, Buddhi and Atman are in the higher or spiritual ternary and are believed not to be experienced by the physical senses at all. The two ternaries are said to be connected by the Ego or I. Some, like Blavatsky in *The Secret Doctrine,* use the term quaternary, which includes the Ego or I in the lower ternary. The New Age view, which is slightly different, sees the seven planes of existence as Physical, Astral, Mental, Buddhic, Atmic, Anupadaka and Adi. The lower quaternary is the realm of the soul and spirit resides in the upper ternary.

The 19th century founder of theosophy, Blavatsky did much to spread Eastern religious philosophy and occult concepts throughout the Western world. Her works, *Isis Unveiled* (1877) and *The Secret Doctrine* (1888), are voluminous and her grammar difficult. Even so, her books are sometimes referred to as the biblical textbooks of esoteric knowledge and used extensively by seekers of higher spiritual insights. She maintained that her doctrines were passed on to her by Oriental religious teachers who attained higher planes of existence. In my opinion, this is an exaggeration to impress and sometimes her work lacks credibility since it cannot be verified. However, there are certain parts of her philosophy that do have merit and are worthy of investigation in the pursuit of spirituality and the soul's nature.

Blavatsky proposes three principles from which other truths can be derived. The first is Brahman which is infinite, absolute, unknowable and ineffectual to speculation. In Hinduism, it is the omnipresent reality believed to penetrate everything in the entire universe consisting of knowledge and bliss incapable of being characterized or defined by any other conscious being. Both spirit and matter are two characteristics or ways in which this deity can give expression. Incorporated with this Brahman is an eternal universe that has infinite dimensions or planes within which a numberless amount of universes periodically manifest. In this boundless playground, there is a process of a constant law in which Blavatsky says there are ceaseless cycles of spirit descending into matter and matter ascending into spirit. Lastly, she believes that all souls are fundamentally similar to a universal over-soul that is an aspect of the Brahman.

Even though all souls are the same in essence, they all differ in degrees of development. Based on the works of Blavatsky, theosophy accordingly teaches seven principles or aspects to human nature. Unlike other living entities on Earth, human beings are said to be complex and advanced beings with both a higher and lower nature. The higher, comprising mind, soul and spirit, has been polluted by the lower physical nature and must be purified before it can return completely to the divine. Purification takes place through a series of incarnations in accordance with cyclic and karmic law.

The densest and most earthly form of our human nature is the physical or biological body made of matter, which is the vehicle of all the other higher principles during life. Linking this body with the inner planes or higher forms of being are the life force and the astral double. The life force is the vital principle that filters through and gives life-like qualities to the physical body. This vital principle or Prana ("chi" in some Chinese traditions) is also referred to as the breath of life. The astral double is a more etheric counterpart of the physical body. The astral nature, which forms the mould for the physical body, is the less dense spiritual counterpart of the physical body that has similar corresponding organs. It even has similar senses to our physical nature that form the illusive links between the physical and spiritual natures. Next, we have the desire or emotional vehicle, which all of us have during a lifetime but does not become a distinct body until after death. Being mostly animalistic, it is considered the seat of our feelings that drive us to action. These four aspects together are called the lower or mortal quaternary of human nature.

Distinguished from the mortal quaternary is the immortal spiritual entity or upper imperishable triad. The three upper principles called the Ego or Higher Self are considered the immortal part of our being. The first of the three is the mind principle or consciousness, also called Manas, which links the higher individuality with the lower personality. It can gravitate downward to the animal nature and identify with the lower nature, or it may gravitate upward towards the spiritual nature and thus attain immortality. Manas is our intelligent faculty and also the self-awareness factor of "I" that distinguishes us from mere animals. The sixth principle, also called Buddhi or spiritual soul, is a universal principle and is not entirely regarded as a human one except for its ability to merge with the universal spirit. The Universal Spirit or Higher Self, also called Atma, is the manifestation of all power and consciousness in our Cosmos. Like Buddhi, the Atma is not regarded as an entirely human principle, but without it no living being can exist.

Blavatsky believed "Spirit" existed first and then through a series of "rounds" when matter was formed, it was eventually able to transfer its nature to matter and manifest itself at the height of the material personification in our present human form. That is, combined with the spiritual nature that informs it. This, however, is made possible through what she called Monad.

For Blavatsky the Monad or "Jîva" has worked its way into the lowest form of matter or the "mineral" body by the Law of Evolution. This, however, happened over millions of years. It first had to pass through stages or "rounds", some of which are stone, vegetable matter, animal matter and eventually the physical man. Up to the so-called "third round" everything was still formless as matter and without consciousness. For Monad to express itself from a spiritual nature to our physical realm, it requires material shape such as our own human form and an intelligent consciousness to guide it. However, this is not possible for senseless living matter and thus, according to Blavatsky, Adam required a soul. She maintains it is also at the latter stages of development that humans became separated into male and female and only then were we endowed with a conscious soul (Manas) that is rational and individualistic.

Gottfried Leibniz, whose works also encompassed theology, law, diplomacy, politics, history, linguistics and physics, introduced the concept of the Monad earlier in the 17th century. He developed a philosophy that combined the mathematical and physical breakthroughs of his time with ancient and medieval religious concepts. For Leibniz the universe consisted of an infinite number of conscious centres or small units of spiritual force or energy, called Monads. All organic matter gets its life from God through this Monad and God is believed to be the Monad of Monads.

The problem with the views expressed above is that they do not really explain the place of the soul well enough – they do not afford it as prominent a place as we require to establish its authenticity. At this point it is merely a differentiation between the physical nature/lower ternary and higher ternary which is spiritual or has to do with spirit rather than soul. I think we require a far better understanding and interpretation of what the soul is or where it fits into the composition of life and human existence.

Geoffrey Farthing, a lifelong student of Theosophy, who held numerous positions in the Theosophical Society in England including General Secretary (1969-72), elaborates on the Theosophy of Monads. In his book, *Theosophy, What's it all about?* (1967) he explains how Monads, after vast periods of time, eventually had the ability to become fully developed humans with physical bodies As this happened, we became more conscious on the particular level on which we happened to be operating. For Farthing, it is the process through which the Ego is able to take possession of a body at birth and then discard it at death. This is similar to earlier chapters dealing with the possibility of parasitic spiritual invasion and possession, a doctrine that I definitely do not support. Farthing also explains the processes of dying where the person dies of causes other than being killed. In such a case, life begins to withdraw from the body beginning at the extremities. It then withdraws slowly into the head until apparently the soul leaves the physical body.

Farthing comments on the lack of consensus about the Monad. Some say it is applicable to the last three principles taken as representing the unity of the spiritual ternary Atma-Buddhi-Manas, "*that immortal part of man which reincarnates in the lower kingdoms*". Others believe the Monad consists only of the last two principles of Atma, pure spirit (ineffective by itself) and its vehicle Buddhi, termed the Spiritual Soul. His interpretation is that modern occultism sees the human Monad as the indestructible element that continues through the cycles of reincarnation, which is present in every individual human. The Monad according to Farthing is said to be the essential unit of the active universal life, which, jointly with Manas, becomes the conscious reincarnating Ego.

After millions of years of this kind of development, we find ourselves in the human body as we know it today. Theosophy therefore teaches that some form of spiritual nature's search for life came first and, once Darwinian evolution reached a stage of maturity, the spiritual Monad was able to express itself fully and even do so consciously. There is, however, absolutely no proof of this theory, even though it is almost a carbon copy of what Newton's hypnotised patients express. Nothing about these theosophical doctrines or the views of Newton's patients is original as they can be traced back to philosophies dating from long before the occult works of even Blavatsky.

In occultism, there is the possibility of a further adaptation for conscious life such as our own. The physical bodies of humans are said to be connected to their astral bodies via a silver cord. This concept of some form of soul being on the other side of the silver cord makes for some serious controversial reading (see Chapter 12). If that is the case, we imply that the soul is linked by a silver cord that can actually be seen by clairvoyants, but if it can be seen and is linked to the spiritual bodies, the soul itself must also be physical in nature and should be visible as well.

This brings us back to the highly contestable and contentious principle of whether the soul is material or immaterial. The materiality of the soul is often used too loosely. The soul cannot be material and immaterial at the same time. What is really meant is that the astral counterpart can be separated from the body and that the astral body is found in suspension outside the body and not the soul. The soul is something else and not the astral double of the self. The question of whether the soul is in suspension and accompanied in the astral body is also still to be explored. The soul, in most occult circles and in my opinion, is immaterial and therefore even if it is with the astral body at the time of separation must still be something other than the astral spiritual counterpart.

It is believed that there is no possibility of the soul re-entering the body once the silver cord has been broken. Farthing continues by saying that once the silver cord is broken, the astral double is freed but soon disintegrates as its vital principle leaves and dissipates. I must say that if all explanations on death and the process thereof stopped here I might have thought it had a measure of credibility, but Farthing goes further. Much of his remaining comments are based purely on speculation. His views that the individual wins and the post-mortem process continues if there is enough spiritual content in the life just lived cannot be verified. Further speculation on the life withdrawing and leaving a dormant psychic corpse behind after a long period within the "mento-emotional" vehicles also cannot be verified. After a complete withdrawal, a second death occurs freeing the individual from the last reminiscence of its personality. All that is left for the individual is but a triplicate vehicle of being, which is the higher thought principle, its intuitional vehicle and its purely spiritual principle.

Farthing writes that whilst there is a disintegration period of the earthly personality vehicles after death, the remainder of the spiritual content of the preceded life is extracted and absorbed by the higher mind principle, which is said to be the spiritual Ego. It was believed that this "distillation" accumulated from past lives give the Ego its individualistic qualities. This is a counter-argument for geneticists who explore the way in which genes differ and mutate. After a period of hundreds of Earth years' existence as a purely Egoic entity, a process of quickening is believed to take place whereby the Ego again gathers "mento-emotional" materials around itself through magnetic means. When the right kind of physical body to suit its needs and fulfil its future destiny is ready, it attaches itself to an embryo body through which it is eventually born. Somehow at this point the soul principles make their reappearance by taking possession of the newly found body and the Ego can live another earthly life. I find this altogether far-fetched and unfounded. I cannot for one moment accept such speculation.

Something to consider, though, is the consciousness that evolves from the belief in the "next world", which Farthing says is for most a world of effects only, and a subjective mental state. By developing personal power and retaining consciousness on the "inner planes" it is possible for humans to develop their astral "senses" and by doing this we can become free-acting entities of this realm. This is said to be a relatively advanced stage depending on which level the individual is active. To me this indicates that humans are currently at a place where they are already assuming the role of conqueror of such alternative realms. We have come to a place where we think we are ready and capable of becoming masters of the spiritual realm and believe it can be a conscious process.

A view many people find quite persuasive is that of the Austrian philosopher and scientist Rudolf Steiner. Originally a theosophist, he broke away in 1912 and founded the Anthroposophical Society. In 1913, he established a school of spiritual science called the Goetheanum at Dornach, Switzerland. Anthroposophy or spiritual science according to Steiner is also synonymous with "occult science" or can be translated as "supersensible" knowledge. His occult view goes into much detail on the seven planes of human nature and central to this is a tripartite nature of the soul. What follows is mostly taken from Steiner's work on the soul from his occult-scientific point of view.

Of all the esoteric theories of the makeup of human nature, I would say Steiner presents one of the most significant, comprehensive and well thought out explanations of what or where the soul nature fits in, assuming we have one in the first place. This then is possibly one of the best attempts to create a blueprint for human nature explaining the nature of our soul from an occult viewpoint. The reason I say if it exists at all, is because even though Steiner at the time gave a credible explanation, his views can in many ways be refuted by current neuroscientific discoveries explaining how the mind is capable of producing an "illusion" of that which the occultists accredit to our spiritual nature.

Physical nature

With similarity to eastern (oriental) and esoteric disciplines, Steiner defines our human nature firstly as having a ***physical body*** consisting of solids, liquids and gasses. The physical body, which he calls mineral in nature, is caught up in the physical realm or world. In the light of occult science, as Steiner calls it, humans appear to be composed of several members. The more physical part or bodily nature as opposed to the pure spiritual component consists of a physical body, etheric body and astral body. In the middle, we have the Ego through which the soul sheds its light. The soul consists of the sentient soul, intellectual soul and spiritual soul. Lastly, we have the spiritual members of Spirit-Self, Life-Spirit and Spirit-Man.

The ***etheric body*** or life-giving force is still part of physical nature, but higher in vibration. Steiner says this etheric body, which is still invisible to ordinary vision, has formative forces that suspend the physical body in a condition we call life. Without this life-giving power, the physical body will disintegrate into its fundamental particles such as atoms. The physical body returns to the mineral world at death when the etheric body is no longer there to support it. The etheric body is said to completely permeate the physical aspect and all the organs of the physical body are maintained in their living form by the etheric body. Every part or organ of the physical body has an etheric replica in the less dense etheric realm which is seen by clairvoyants as a pale, luminescent vibration extending about 15 mm around the physical body and has a blue-grey to violet-grey colour. Steiner was not clear whether this etheric double could separate from the rest of the body as is said to be the case with astral projection. It is something we have yet to explore.

The next structure is the ***astral body***, which is the spiritual counterpart of the etheric double and also the informing or life-giving force whilst the body is still alive. The astral body, which can also be seen through clairvoyance, is even higher in vibration than the etheric body and called the spiritual body even though still considered part of the physical body. It is this part of our physical nature that can function outside the confines of the physical and etheric bodies. Steiner uses the astral double to explain the "ghost", which people sometimes see shortly after someone dies. Steiner firmly believed that the astral body is the bearer of pleasure, pain, joy and suffering, and that during sleep it separates from the etheric body or is lifted out of it until the sleeper awakes. None of the above is the soul. The soul comes next and it becomes clearer with Steiner's views that it has a mental and therefore completely non-physical nature.

The associations made between the etheric body and the astral body as replicas of the physical body as they also form the life-giving forces of the bodies lower in vibration, becomes very important later on when I deal with genetic manipulation to recreate life through altering our DNA structures. A factor overlooked in the field of genetics is the etheric and astral DNA counterparts that might exist if we do have etheric and astral bodies. If the etheric and astral bodies are merely higher in vibration, should we not perhaps create technology and know-how to observe and study these subtle bodies before trying to recreate our nature through genetic means related to the physical realm alone?

We are finally presented with the ***Ego*** or "**I**" which, Steiner says, is the reincarnating vehicle of consciousness and the first time we encounter the actual soul part of our nature. He envisions the ego as the reincarnating vehicle of consciousness living within the sheaths of all three parts: the physical, etheric and astral. This fourth member is the unique element that distinguishes human beings from any other known living creature in the universe. Steiner has a unique tripartite view on the soul nature consisting of the *Sentient Soul, Intellectual Soul* and *Spiritual Soul.*

Soul nature

The ***sentient soul*** means exactly that, a soul interacting with the physical senses. While you perceive an object, a mental image or picture is formed. By turning away, you take sensory perception away but you are still able to recall an image since the memory is already formed. According to Steiner a process was taking place between the astral body and the ego whereby the astral body brought the external impression of the object to the observer's consciousness. For this awareness of the object to last, Steiner says the ego is required to receive this awareness into itself and make it its own. It is at this point that Steiner distinguishes "body" from "soul". His hypothesis is that we speak of the "astral body" whilst we have in mind how the knowledge or awareness of an object comes about, but it is the "soul" that gives the knowledge performance and duration. From this, he says, it is evident how close the connection of the astral body is to the part of the soul that gives permanence to knowledge. The closeness of the connection between astral and sentient soul gives the impression the two are united, making it possible to refer to them jointly as the astral body. Steiner calls the astral body of man the "Soul-Body" and as far as soul is united with the astral body, the "Sentient Soul".

With the ***intellectual soul*** the ego further advances itself by becoming conscious of the external objects. In order to free ourselves progressively from the external objects and become independent to work within our own faculty, we require what Steiner calls the "Intellectual" or "Mind-Soul". Both the sentient and intellectual soul work with the information received through sensory perceptions and the memory retained of these impressions. It is also at this point that humans developed the ability to distinguish between external objects and the internal "I", which distinguishes us from external objects. No life form can attribute to itself the knowledge or consciousness of "I" without the ability to identify with itself in such a manner. Only humans, Steiner points out, have developed the characteristics needed to apply this persona to the personal self.

The highest form or ***spiritual soul***, on the other hand, is attained when the soul becomes aware of its own fundamental nature. It is also this third member that becomes part of the Divine. Steiner does not imply that the "I" is God, but only that its essence can be likened to the Divine. The third member of the soul gives rise to the inner knowledge and awareness (consciousness) of our nature as living beings, which is distinct from the astral body giving us the awareness of the world outside of our being. This third member of the soul Steiner calls the "Consciousness-Soul" or "Spiritual Soul".

Having realised the physical part of the physical body, with esoteric and astral bodies that exist on a higher vibratory plane, Steiner explains how humans may experience an even higher order of spiritual nature which is altogether non-material but similar since it is also tripartite in nature. The upper ternary is presented by Steiner as the "*Spirit Self*", the "*Life Spirit*" and "*Spirit Man*". The ego or "I" links the lower and higher ternaries through the soul. Not everybody experiences consciousness of their higher spiritual being and neither do all achieve full development and maturity thereof. For Steiner the higher an individual develops on the three spiritual levels the more he or she is able to become master of the lower physical bodies. Once this is achieved, a reverse process is set in motion whereby the more advanced spiritual natures have an effect on the lower of the physical bodies respectively in descending order. Steiner believes by doing this we extend and master our astral, ethereal and then physical natures. However, the process seems to occur from the top down rather, as one would assume, from the lower physical nature working upwards towards the higher astral plane. The most important factor to keep in mind is that the ego is the mediating factor for the spiritual development that follows. The Ego or "I" makes humans an entirely unique species considered the only beings capable of developing the higher spiritual faculties.

Spirit Nature

The lower of the three spiritual bodies, considered to be the most important faculty which allows higher levels of evolutionary growth and development, is the ***spirit self***. Steiner points out that spirit self is synonymous with the term Manas of oriental classification. By coming to understand the hidden forces that underlie the astral body, we can advance ourselves higher in spiritual nature. As we advance ourselves, the Ego firstly instructs and forms an amalgamation with the astral body. It is this transformative process between astral body and "I" that Steiner calls the spirit self. As we continue to work upon ourselves through the spirit self, it manifests itself in a higher maturity of our feelings and will.

The second fundamental interaction humans have in their quest for enlightenment is between the Ego and the etheric body which Steiner calls ***life spirit***, also known as Buddhi in oriental and esoteric disciplines. Somewhat more intense and demanding, the work of the Ego on the etheric body is towards transformation and takes place irrespective of the individual's conscious awareness. According to Steiner, a person's life experiences and religion become engraved in his etheric body, changing it into life-spirit. It influences the etheric body with time and especially through constant repetition. For most it is more or less an unconscious process, but through initiation in what Steiner calls supersensible knowledge, which may represent enlightenment in other circles, one may be able to become fully conscious of this activity.

Steiner ends his hierarchical arrangement with the ***spirit man***, also called Atma, which constitutes the interaction of the Ego with the physical body. Most people are completely unaware of this process and even those who are conscious thereof, are very dimly so. Full enlightenment on this level is therefore only reached with supreme spiritual knowledge and insight. In order to unite the Ego with this hidden Spirit, the highest member, the spirit man is required to work upon the lowest or physical body of human nature.

As we evolve and develop our souls, our spiritual awakening increases and the human emotions, desires and passions decrease. Steiner likens the human soul of an unenlightened person and a spiritually evolved one, to a pyramid and an inverted pyramid. In the case of an unenlightened soul, human passion, desires and emotions form the broader base of the pyramid. The smallest part or apex would be representative of the person's aptitude on the level of superior spiritual consciousness. The opposite is true for the spiritually evolved soul represented by the inverted pyramid, where the smaller apex at the bottom represents the worldly desires, emotions and passions, and the broader base at the top the enormously broadened enlightenment and high degree of cosmic consciousness.

The Occult Contradiction

For a long time now occultism has presented us with what seems to be a credible explanation for the multi-faceted nature and supernatural behaviour humans sometimes display. This reflection is particularly pertinent when we cannot explain uncanny behaviour and ability through our current scientific knowledge of human nature. With superstition still prevalent, supernatural abilities are judged as paranormal behaviour because they cannot be explained. Due to persecution and harsh punishment sometimes resulting in death penalties, occultist explorations were pursued in secret. With the church and other orthodox bastions losing their grip in modern civilizations, more of these practitioners came to the fore and presented the world with extraordinary explanations of some of these phenomena. Some are based on old historic tradition and some on newly found psychic and clarient experience, but unfortunately others are truly nothing more than mere speculation. It is therefore still difficult to discern between truth and fantasy.

The broad occultist view of the soul still contains the same paradoxes as the well-debated philosophies such as dualism and Cartesian dualism. Current neurobiological discoveries are also starting to cast doubt on the existence of alternative spiritual personas, since all can be attributed to the biology of heredity and variation in organisms seen in genetics. Occultism is unable to establish how and at what point interaction between soul and matter takes place. Nobody has ever been able to successfully examine and explain, let alone prove, the point of amalgamation without reverting to theory or belief. Even with clairvoyant testimony of alternative realities and spiritual bodies, one could similarly explain all existence through modern neurobiology and the neurosciences. According to these disciplines all characteristics of life are locked up in a message carried in the DNA and all mind processes and cognition can be explained as products of the brain.

We are far from developing technology that is able to reveal these alternative, less dense physical bodies such as the etheric and astral in any way, let alone conclusively. For the soul to exist and to have no physical or material nature whatsoever is difficult to comprehend, since it seems impossible for anything to exist without some sort of body or at least energetic form. This is why so many people try to equate soul with some form of ghostly or angelic image. It makes it more tangible and easier to relate to. However, the moment we attribute form or energy to it we no longer have an immaterial soul. In many religions and philosophies soul is spoken of as an immaterial element. Immaterial implies exactly that: it has no mass and therefore no energy either, which in turn cannot have any form that is visible to any of our five physical senses or any technology available today.

Occultism, in its view of trichotomy, implies that the soul is riding on the back of something else. The soul is seen as the mediator between spirit and matter. During physical life on Earth, the soul requires a physio-biological body. It cannot exist without the help of an existing life form. It needs the physical body either "alive", such as a live embryo, or otherwise it needs an astral body when it wants to wander off outside the confines of the physical body. In this model there is no reason to think that it actually has the ability to exist without such hosts, making it a parasitic union. In the occult domain the soul does not have the ability to stand alone in the physical universe, but requires a host. The soul in this context is a "parasite".

The parasitic union of the soul with a physical body persona interests me. For the soul to exist, it seems logical that it must have such a dependent relationship. The material body plays host to the soul for as long as it takes to accomplish its symbiotic relationship on Earth. The moment the physical body and less dense ethereal and astral bodies die, the soul most likely still dies with it. We have no proof that the prospective soul can survive the finality of physiological death yet. It does, however, allow us to speculate about the possibility that it may one day achieve this goal.

Chapter 11

Materialism and Scientific Model

How does spirituality and the soul stand up to the scientific and philosophical models of today? Simply, it does not. If we probe deeper into the new scientific model, we see it does not provide the luxury of free will either.

Although we have come a long way in understanding the human brain and accompanying mind, we still cannot explain how this incredible organ is capable of achieving consciousness reflecting emotions, reason, thought, desire and retaining memory. As much as there is criticism against the Cartesian model's inability to link the soul with the conscious process, science and philosophy can be criticized for being unable to validate if the mind is physical or non-physical. Simply put, our modern day scientists do not entirely know how the mind works, let alone the soul if there is one. However, present day philosophical as well as neurobiological views are strongly influenced by the materialistic model of human nature. Susan Blackmore in *The Meme Machine* writes there is no room in neurobiology for the soul. According to her, the more understanding we have about how neurons and synapses work, the less need there is for a ghost to control the machine.

Materialism in philosophical context explains that all existence relates to matter or is a characteristic or result of matter. This school of thought teaches that matter is the elemental component of reality, and physio-chemical changes in the nervous system of the brain are the ultimate source of consciousness. Since Charles Darwin's theory of evolution in the 19th century, the philosophical model of materialism has largely been adapted to accommodate his evolutionary principles. Materialism is antonymous with idealism in which the primacy of the mind is affirmed and matter is defined as an aspect or merely an idea of the mind.

Many philosophers today believe materialism will ultimately be successful where idealism (all reality is but an idea) or even neutral monism (perceived reality is neither exclusively mental nor solely physical) seems to fail. In monism, only one basic substance exists. All experience of life can either be attributed to purely mental activity represented in idealistic monism or be purely physical as held by materialistic monism. Materialism, however, comes in many different varieties and each faces formidable difficulties, the foremost being that it is almost impossible for most people to discard their spiritual experience and put it down to a mere illusion of the mind.

Most cultures believe in the existence of a soul or spirit of which almost half believe the soul can separate from the body (*The Meme Machine,* Blackmore – *Shells* 1978). People seem to think that the soul presents itself as the inner self that can survive when the body dies. Nevertheless, as Blackmore rightfully points out, we are unable to find "separate stuff" we can call the soul, and if someone had to it would be incorporated as physical and no longer be separate at all. She also maintains that if it cannot be found by any physical measure it is near impossible to see how it could control the brain.

Owen Flanagan, professor of philosophy at Duke University in Durham, North Carolina and author of *The Problem of the Soul*, deals with the conflict between the humanistic and the scientific – the two grand images of who we are. Flanagan explains that in the humanistic image humans are seen as spiritual beings empowered with free will that permits us to outsmart ordinary laws of cause and effect, a capability no other ordinary animal possesses. The scientific image is quite different in that it says humans are actually animals that have merely evolved according to the principles of natural selection. Although we are extraordinary animals, we do not have the capacity to beat the laws of cause and effect. The big question is which one it is – humanistic or scientific. They radically oppose each other and are incompatible, yet both could be right. Humanism holds that humans contain a spiritual part with an incorporeal mind (or soul), with life and eternal fate determined by the state of this soul. The other is adamant that we are nothing more than social animals and ultimately mortal since we have no soul and therefore eternal life after biological death can not be dependent on the state or existence of such a soul.

The concept of the humanistic image includes a non-physical mind with a free will and a permanent self or lasting soul that is changeless. This view is completely rejected in mental and human sciences and the soul is thought an old-fashioned term rarely used unless in a religious context and even then only by true believers. Flanagan explains that the problem with the humanistic view is that for humans to become immortal, the philosophy relies entirely on an incorporeal soul or mind, and that is exactly why the soul itself presents a problem.

In elaborating the problem, Flanagan looks at the concept of free will in greater depth. If the non-physical mind, free will and the soul are not real but mere appearances, then it is the end of the world as we know it and Descartes' error remains influential. Flanagan proposes that the humanistic view presents two main mistakes: firstly it claims that our minds are able to be disembodied and secondly that our wills can be free. The question posed is how a non-physical mind can cause a body to move. If the mind is incorporeal or immaterial, how can we receive information about the external world through our senses? How do we cross that line that divides the physical and the mental so that we consciously perceive things outside us such as material objects in creation? The Cartesian idea of free will relies entirely on the concept that mental events and choices cause bodily events. If we reject the idea that the mental and physical can actually interact, we cannot explain how it happens since we have yet to prove this scientifically. We similarly have to reject the Cartesian definition of free will since it relies on that very fundamental principle. In the Cartesian view, I choose to perform some action and do so of my own free will, meaning that I choose to act in a way that is not subject to ordinary causal law but by the command of my mind (or soul for that matter).

As explained in a previous chapter, this problem was originally overcome by fabricating a philosophy whereby God created the universe with a set plan for each person's mind and body to run in parallel (according to Leibniz). According to Malebranche, on each and every occasion where there seems to be causal interaction between mind and body, there is no such interaction and it is God who causes the mind and the body to behave harmoniously. Flanagan points out that in neither view do mental events and bodily events causally interact, and neither requires causation even within the mental and bodily realms.

Parallelism is compatible with determinism in that it denies that humans have a free will. For Flanagan neither Leibniz's nor Malebranche's view is possible and even if we reject Cartesian dualism and accept a naturalistic representation of mind, free will remains undetermined until we fully understand the nature and types of causation that exist when it comes to the types that govern mind. Flanagan says the mind/brain might work in a way still to be determined that is not deterministic and that resembles something like the Cartesian idea of free will. This is rejected outright by the scientific model, according to which we are all animal and the brain is our soul.

Flanagan's points derive from the scientific image that the mind or the soul is the brain and is motivated by advances in evolutionary biology, genetics, the sciences of the mind, cognitive science and cognitive neuroscience. Consciousness, cognition, and volition are perfectly natural abilities of humans involved with complex commerce in the natural and social environments. In this context, we possess no special alternative capabilities or ingredients that we know of that can do the work of the mind, the soul or free will. "Desouling" or letting go of the belief in the concept of humans having a soul is a minimal requirement in the scientific image. In the scientific model, it becomes compulsory to discard the soul or non-physical minds. Flanagan is adamant that if such things did exist, as perennial philosophy conceives them, science would be unable to explain people. However, they do not exist, therefore people can be explained scientifically.

Furthermore, we will need to think of people as natural creatures completely obedient and responsive to natural law, which is part of our physical nature. Flanagan rightly points out that the set of religious beliefs whereby humans are positioned between animals on one side and angels and God on the other, needs to be replaced. Although we are complex and unusual animals, we are merely animals nonetheless. This means that the scientific image is committed to desouling us.

Flanagan agrees with science that there is no spiritual component as traditionally conceived, but that we rather make room for human perfection within the constraints of the scientific image. He argues that we now live in the age of mind-science in which cognitive science and cognitive neuroscience use reliable methods and tools to examine and identify the way the mind/brain works. The working brain combined with the rest of the nervous system is our thinking stuff. But spiritual forces do not have any explanations. Unless God and the angels are interfering with us or unless astral forces and extraterrestrials have an influence on our brains and perceptions, we will find that science will one day provide a true picture of our Dasein. (The term Dasein comes from the German philosopher Heidegger, who questioned what kind of being or "Sein" human beings have and called it a kind of Dasein, which literally means "existence" or "being-there"). Mind-science in concert with the other human sciences is gradually overturning the traditional conception of mind. The scientific image concludes we are conscious animals living at a certain time in the broad sweep of natural and social history.

Flanagan explains that there are two assumptions in the scientific image. Firstly that the mind is the brain and secondly that the mind is obedient to natural law, which is possibly strictly causal. In the first instance the mental events can be said to be brain events or, in a more specific light, we can say that mental events are central nervous system events or even whole-person events. The two assumptions have become a given in modern mind-science even though they possibly still offend traditional views. The idea of a physical mind that is not incorporeal and which has evolved according to Darwinian evolutionary principles, has reached the status of truth within mind-science. In this image, free will has no place in mind-science. The question from the humanistic perspective becomes whether the scientific image can be seen as depressing or inhospitable to a dignified, moral and meaningful life, and whether it is dehumanizing as it drains life of meaning. It can appear that life has no transcendent purpose and moral living becomes one among several "quirky features" of our kind of animal.

Flanagan is of the view that this new and somewhat unusual situation, without a soul, is not all that bad. But he also points out that no advocate of the scientific image has ever been able to adequately explain and explicitly show how this image can deny the soul. Both images have the same goal in mind – to maintain a strong conception of what it means to be a person with consciousness, capacities for self-knowledge and the ability to live rationally, morally and meaningfully. Flanagan is of the notion that we can do without all the concepts that are central to the humanistic image in its present form, whereby the soul reigns supreme and still retains some or most of what these concepts were designed to do. Even without souls, non-physical minds, immutable selves or free will, there are still people.

Where the humanistic image displays our spiritual nature, science discovers and displays the external world and our animal essence. That image has now become one where humans are conscious social animals that deliberate, reason and make choices. We possess an evolving or continuous identity that seeks to live morally and meaningfully. With the scientific image, we lose the cluster of philosophical concepts that has been at the centre of the humanistic image.

Flanagan's view is that these concepts do not refer to anything real and we are therefore best off without them. He goes as far as to say that once they are eliminated we still have a picture of people as rational social animals, which is more or less what most people desire. Mind, morals and the meaning of life are all still in place and supernatural concepts have no actual philosophical merit. There simply are no good arguments from a theological, philosophical, humanistic or scientific point of view for belief in divine beings, miracles, or heavenly afterlives.

Flanagan points out that the scientific view is still seen as negative since it has not yet really been able to show the difference between what it means to be an animal and what it means to be a person. Even though we know we are human animals, we do not know how to tell the story without it being alarming. For many, Darwinian evolution renders life meaningless and it removes purpose from the universe, making it difficult to accept that we are animals. At the heart of this thought lies nihilism with its concept that something without purpose has no meaning.

Lastly, Flanagan holds the view that Darwinism showed how the evolution of species is unplanned, but the theory of evolution by natural selection does not imply that life has no meaning or purpose. Even though evolution produced intelligence, it does not require intelligence to bring about intelligence. Evolution also produced meaning. We have language, words and signs, which all have meaning. It is a common mistake to think that evolution drains the universe of all meaning and purpose. Evolution is a process through which life-forms, behaviours and consciousness emerge. There is no need for a creator God outside the universe giving it meaning and purpose, and that includes a heavenly afterlife once our animal physical part dies. This idea will have to go, since our animal part is our only part and once we die, we no longer exist.

Flanagan goes on to explain that it is not so bad after all for us to be only animals, but I will not explore the rest of his philosophy. Not a pretty picture from a humanistic and religious point of view, but that is what the new scientific model holds. In fact, it is not a great place to find ourselves from any point of view. The soul is dead and any idea of it has been reduced to nothing more than an illusion or fantasy for all of human conscious life thus far. A figment of our imagination, myth and mythology, magic and ritual still upheld by the stupidity and ignorance of billions who refuse to come to terms with our mere animality. Can this really be true? Is all OK without a soul?

PART 3

Homo Angelicansis

Chapter 12

Triunism (The new -ism)

No! All is not okay without a soul. Philosophising the soul into oblivion with the help of Darwinism, materialism or any other -ism reduces our experience to mere fantasy or illusion. For too long we have lived with the immensely deep-rooted impression that a soul does exist. The idea of a soul is deeply implanted in our daily experience of life. Unexplained "supernatural" powers displayed by humans clearly indicate that there is more to our nature than mere organic matter somehow acquiring life and consciousness. Yet we argue whether the soul exists and plays a role as a life-giving force or adds some value to life. I have shown how our modern scientific models reject the notion of the soul and maintain that all mind, memory and consciousness are simply products of the brain. Religion offers us belief and dogma to substantiate the soul as the life-giving force. Science and religion are in direct conflict, yet both have merit.

On the one hand, we have humanism, dualism, occultism and religious dogma with the absolute conviction that life and mind is linked to the soul. On the other hand, we have scientific models insisting we are physio-biological beings with "physical" minds only (the mind is the product of the physical brain). For these disciplines, the soul is not a prerequisite for the continuance of life, mind and consciousness of any kind. Regardless of this view, the belief in a soul will simply not go away. The soul refuses to die; pun intended. Even though science has discarded the soul, it remains an integral part of the belief systems of billions of people worldwide. Is there a soul, or are all those who think so, deluded? In a similar fashion to the religions constituting their faith on potentially fictitious "gods" in the flesh, perhaps the majority of the world's population also believe in a potentially fictitious entity we call the soul.

Religions rely on conditioning and indoctrination for their success. However, faith in any doctrine has no purpose. We require knowledge. We are able to gain such knowledge and insight through constant investigation. Presently, the soul is still an unknown factor. With persistent probing, we should be able to verify its existence. If there is a soul, we must examine its nature and discover its purpose. Currently our best estimation determines that the soul is the vehicle of at least part of our consciousness beyond physio-biological death.

It would be impossible to include a critique of everything ever said and written about the soul. My earlier chapters detailing the history of human perception regarding the soul are comprehensive enough. They show how it is impossible to affirm some of the views and how others are not worth the time of day. Likewise, some of the most established doctrines incorporating the soul have never been challenged successfully and I will once again attempt to show how little we really know about the human soul. There are a few conjectures that stand out enough to demand further investigation.

There are many philosophical, religious dogmatic and esoteric –isms asserting the notion of a spiritual realm with a soul-mind duality. This contrasts with materialistic and scientific models maintaining there is no spiritual reality, for all life consists of matter and mind only. So, which -ism is the right one? Which philosopher, scientist, poet, cleric, prophet, messiah, disciple, medium, clairvoyant, occultist, shaman or the like is right and which doctrine, religion, philosophy and/or theory is the one that contains the truth about death and the soul? How ironic that there is a strong probability we will never know, since on the one hand there may not be an ultimate truth, and on the other we may simply run out of time before we have the chance to find out.

Over the years our perception of the soul has changed and even today we have many differing views on whether the soul exists and what it is. The answer to our changing perception is twofold. What we thought of spirituality and the soul thousands of years ago becomes obsolete as we learn more about our environment and ourselves. New knowledge and insight forces us to constantly upgrade older perceptions. However, it could also mean that the soul itself is changing and as it changes, our perception must adapt to a new form of evolutionary soul.

According to the new scientific model, the mind is physical and merely the product of the brain. There is therefore no room for a secondary informative soul function. Perhaps the brain-mind is physical by nature as our neurosciences suggest. However, this does not necessarily mean that a non-physical (immaterial) soul does not exist. Perhaps the neurosciences and Cartesian Dualism are wrong to liken the rational mind with soul. When you look carefully at what the Book of Matthew and occultism says, you will notice that both separate the brain-mind from the soul-mind. There are two minds! Two minds; what are we talking about? Where did the other mind come from?

I think it is time to re-assess some of the scientific disciplines before we discard a dualistic and/or trichotomous consideration of human nature. Perhaps we missed something along the way. Perhaps we can come to new insights now that we have managed to overcome the fantasies of our mythical past and the false impressions based on our illusive religious doctrine. I cannot help but see a grey area comprising a missing link that needs to be addressed here. We need to find some kind of unifying theory between spirituality and our physical reality so to speak. The further our scientific knowledge of our cosmos develops and the more we discover about evolution, the further removed we become from recognising and acknowledging our spiritual nature. Yet, that is exactly where I think the answer to our problem lies.

Even after the study of the history, makeup, eschatology and presence of the soul has revealed that it may not actually exist, do we still need to pursue the matter? Darwinism shows that we are merely the most advanced and highly evolved animals. If we are so advanced, at what point did we acquire souls capable of spiritual life, but animals such as our nearest evolutionary cousin, the chimpanzee, did not? Why can a human animal have a soul which is so superior that it is able to achieve salvation after death? Why do we judge humans for their actions when no other animal is scrutinised by the same standard? If we accept the supposition of religious salvation, then we create in our minds a God of redemption and salvation, but at the same time we create a cruel God who reserves immortality exclusively for humans. Seen against the backdrop of life becoming more complex as evolution takes place, we need to ask when humans acquired this elevated status of soul-beings. If we do not accept the fabulous religious mythology of a "God" breathing life into humans to have a soul but not into other animals, what is the alternative?

In order to establish whether human empowerment with a soul is credible, we have to be able to place it in the new scientific model. The soul must fit into the current scientific models dealing with the origin and evolution of the universe and life on Earth. We can no longer accept the fabled story of Adam and Eve created in one fabulous act as prescribed by religion. There is simply overwhelming proof that our universe existed for more than 14 billions of years and life on Earth for more than 3,5 billion years. We also have abundant proof of our own evolution as genus Homo developing for the last 4-6 million years. So, where does the soul fit into the structures of cosmogony and evolution?

Cosmogony, archaeology and palaeontology in many ways contradict our religious and esoteric doctrines. Christianity's breath of a fresh soul into the first man is fantastic, the soul sought by philosophers as the informant of the mind itself is unlikely and the monistic physical mind of the neurosciences is unconvincing. In addition, the soul as a by-product of extraterrestrial activity is outrageous, the soul as a ghost-like figure in alternative vibratory realms has some merit but not without contradictions, and let's rather not mention the preposterous reptilians again. So what then will satisfy me in the quest for truth and a better understanding of the soul? For that, I have to turn to the cosmogony of our universe and evolution of life. The journey takes us from the beginning of time when our universe exploded into being, right up to the forming of life on Earth and the subsequent evolution thereof. In all this time, at what point did the soul come into existence?

The trichotomous doctrine of Catholicism holds that humans consist of body, soul and spirit, whereas Cartesian dualism claims we have a body with a mind that is also the soul. Nevertheless, according to the Book of Matthew, "Jesus" contradicts both these views by claiming human nature consists of body (heart), soul and a mind. Against Jesus, biblical Paul's view of a trichotomous body, soul, spirit unity seems erroneous. Academically, Paul's point of view is incorrect due to the mere fact that he excludes the mind altogether and thus contradicts the view of a supposed Messiah. Paul's conviction of body, soul and spirit is most likely based on the struggling dualistic perspective that the soul is the mind and that the two can therefore be seen as one. This to me is incorrect. It is equally questionable from a Christian point of view since it contradicts Jesus, whom the doctrine teaches is the "Son of your God" and "God in the flesh" (*John 1*). According to the faith, Paul contradicts "God".

The traditional concept of mind-body dualism holds that the soul (psyche) is the mind and that the soul can exist independently and outside of the human body. Modern thinkers reject the notion of dualism. For them the mind is a consequence of brain activity and under no circumstance is the mind separable or able to exist independently from the body (brain). Trichotomy includes a body with a soul and spirit but excludes the mind. The solution I suggest is a triune nature of heart, soul and mind as the Gospel of Matthew credits Jesus with saying. For this triune alliance, I create the name "Triunism".

Triunism is the concept where a living organism has a physio-biological body, a cognitive brain function and an added spiritual component of a separable soul. The cognitive brain function is electro-chemical, resulting from neurological activity carrying electrical charges resulting in chemical interactions as the neurosciences suggest. In addition, we have a soul-mind currently expressed through the subconscious (mind) that is somehow connected to alternative dimensions and realities. And yes, the soul nature can exist outside and perhaps even independently of the physical body. However, we are by no means sure yet whether the soul is immortal. Triunism holds that humans are physical beings in "two minds" so to speak, a state of being caused by eons of evolution.

Even if not proven yet, it is already abundantly clear that the rational mind function is the product of the brain. Neurobiology shows that cognisance and memory stops when the brain dies. Nevertheless, this does not mean there is no soul-mind capable of surviving biological death. There is ample reason to think that we have yet another mind at work with its own consciousness in some respects capable of outperforming the brain-mind. What would constitute proof that there is a "soul-mind" independent of the "brain-mind"? If we can detect human-like consciousness outside the confines of the brain and its mind, we have grounds for soul to exist. It would have to be mind at a distance so to speak. But, this is exactly the reason why we think there is a soul in the first place. We do ostensibly find consciousness outside the confines of the physical body. Human consciousness that is!

In occult science, it is said that the astral body is responsible for making mind and brain interact. Occultists hold that the function of the astral body is threefold. It makes sensation possible, it somehow creates that illusive bridge between mind and physical matter, and as a vehicle realises independence for conscious activity. It seems contradictory that the astral body or "soul-body" as Steiner calls it, forms the bridge between the physical body and mind or that soul is responsible for the mind's consciousness, since it borders on the long disputed view of dualism. This is in stark contrast with scientific models claiming we have a physical mind only. Occult science bases its philosophy on triunism where the soul is a third member independent of physical body and mind.

According to occultism, the astral body leads an independent existence on the astral plane when the physical body is asleep or during astral projection. This idea of travel without the physical body points to the all-important occult teaching that humans may develop the astral body to a level where it may become a vehicle for unbroken consciousness on both the physical as well as the astral plane. There is a possibility that occult phenomena such as astral projection and clairvoyance are connected with the use of the astral body in higher dimensions of reality. Such activity is said to be unconscious in ordinary humans, but can be a completely conscious process for highly developed and enlightened people. The occult belief is that the astral body can leave the physical and etheric bodies behind and function independently on other planes of reality as long as it stays linked to the dense physical body by means of the astral cord or silver cord.

Alexander Cannon in *Sleeping Through Space* noticed the Bible also speaks of the astral body as the "Golden Bowl" and the etheric body as the "Silver Cord". Ecclesiastes 12:6 states: "*Yes remember your Creator now while you are young, before the silver cord of life snaps, and the golden bowl is broken,.....*" (Tyndale – *The Living Bible – Paraphrased*). The King James Version is less expressive: "*or even the silver cord be loosed, or the golden bowl be broken...*" According to Cannon, the astral body surrounds the physical body as an eggshell surrounds the contents within it. However, he falls short of an adequate explanation by alleging the astral body is linked to the physical body by invisible vibrations on the ether in the air and allegedly through the solar plexus. Cannon further alleges that the astral body is mainly the emotional body and has to do with emotions, moods and feelings. The astral body links up with the physical body through the solar plexus and similarly the etheric body through vibration passing from it through the physical body between the eyes to the top of the etheric body.

Astral projection is commonly known as the ability of an operator to consciously project his or her consciousness and visual perception in the form of the astral body outside the confines of the physical body whilst still being linked to the body with what is commonly known as the silver cord. There has been much debate about the vulnerability of the physical body during such projections. Some practitioners are of the opinion that the body is left empty and unguarded during projection. If the soul-mind left the physical body empty during projection, it would be logical to assume that this empty body could be vulnerable to invasion or intrusion. However, the fear is not always of an attack on the physical body from outside, but from an internal spiritual nature such as a psychic attack or even of possession of some spirit or demon. This to me is unfounded and fear-based, set on the same old beliefs in the supernatural and magic and regrettably still fantasy. There are no such supernatural celestials. To me the only fear should be of outside interference, since the physical body is subdued and left without its full conscious function. Return to the physical body can be traumatic if forced to do so by an outside influence.

Bruce Roberts in his book *Astral Dynamics* describes how one can achieve a "mind-split" during astral projection, which is an out of body experience (OBE). Initially it was thought that the body is left alone when a person's astral double exits the physical body, but Roberts thinks that is not the case. I personally agree with him in that some consciousness is still present with the physio-biological entity. Some people say the etheric body exits the physical body and some say it is the astral body. However, if we consider the etheric body as the life-giving force of the physical body as previously explained, it would be difficult to imagine it leaving the physical body without the body dying.

Roberts thinks that the physical body is never left empty while it lives. In his experience, one's consciousness reflects a copy of itself outside the bounds of the physical body while it sleeps or during conscious out of body experiences (OBEs). Roberts thinks the reflection into an energetic vehicle capable of supporting and maintaining its integrity outside the bounds of the physical body is a kind of energetic echo containing a complete copy of consciousness, mind and memory. However, the physical body is never empty and unguarded. While we are still alive, the original or master copy of consciousness never leaves the bounds of the physical body. When the physical body falls asleep or enters the trance state, consciousness is automatically projected into the expanding etheric body. An energetic reflection of consciousness can then exist at a slightly higher energetic state than the physical body is capable of allowing. The etheric body at this point contains within it a full reflection or manifestation of a copy of all memories, conscious and subconscious. The expanded etheric body is now an energetic step above the physical body, which Roberts says is one energetic step closer to the real-time body and the first level of the projectable double.

Roberts goes into much more detail about the state of the etheric body and in his opinion, it stays fully conscious if held in the trance state during a conscious OBE. The etheric body according to Roberts can also function altogether independently from its projected and dream-mind aspects. The physical combined with the etheric body generates the real-time body internally, which becomes the first projectable double. Similar to Steiner he says it is a natural occurrence during sleep, but can also occur during the trance state. As the real-time body is internally generated, another full copy of consciousness and memory is automatically reflected into it. The real-time body's thoughts are fully synchronised with those of the etheric body as long as the reflection of consciousness is still fully enmeshed within the physical/etheric body. Roberts says no noticeable mind-split effect has yet occurred at this point, even though the energetic conditions for a mind-split are already present. Everything is now set for an out of body projection to happen.

At the precise moment of separation when the real-time body projects, the first true mind-split occurs. Roberts says that after projection the real-time body exists completely independent as a freestanding entity capable of thinking, experiencing, and registering memories at its remote location away from its physical/etheric body. Once the projection of the real-time double has taken place, the expanded etheric body that has it own copy of alert consciousness remains cognisant for as long as it can hold itself awake. It is here that the real-time double is capable of operating completely independently from its physical/etheric double and at a distant location outside the physio-biological body. The real-time projected double, he says, will often even feel the physical body and hear sounds coming from that location.

Roberts explains that this process results in two complete energetic copies plus a master copy of consciousness and memory. The master copy is said to be the physical/dream-mind, the expanded etheric body is the physical/etheric mind and the first true mind-split is represented by the real-time projected double. All three copies are energetically interconnected and all three are capable of functioning and thinking completely independently. For Roberts the mind-split is quite evident and consciousness resides in both the physical body with its mind as well as its projected double at the same time.

Medical researchers have been fascinated by the occurrence of so-called near-death experience (NDE). In NDE, people have extraordinary sensations when they are close to dying or even when they felt like they might have died. Professor of psychiatry and a member of the Parapsychological Association Dr Bruce Greyson has focused his research on NDEs for the past thirty years. Greyson gathered empirical data on NDEs using accepted scientific methods to construct the *Weighted Core Experience Index* to help measure the depth of the NDE, and the *Near-Death Experience Scale* to distinguish the genuineness of the NDE. The Greyson scale gives sixteen features such as experiences of an altered state of time, accelerated thought, reviewing of life, a sense of sudden understanding, feelings of peace, joy and cosmic oneness, seeing or feeling surrounded by light, vivid sensations, extrasensory perception, visions, a sense of being out of the physical body, a sense of 'otherworldly' environment, a sense of a mystical entity, a sense of deceased and/or religious figures and a sense of a point of no return.

An article by J. Timothy Green called *Near Death Experiences, Shamanism, and the Scientific Method* documents the near-death phenomenon over the past 20 years. According to Green, shamans can enter into an altered state of consciousness and have an NDE at will. Numerous methods are used to achieve this state, but the one most frequently used is fasting, flora-potions or prolonged drumming. Once having achieved the ecstatic state, the shaman enters into an extraordinary reality that allows travel to one of the three realms for a specific purpose. Green tells us that Shamanic cosmology postulates three different, but interpenetrating realms of existence. These realms come across as real even though they are experienced as completely different from the physical reality of our natural physical environment. For shamans spirits populate these worlds.

Shamans enter into altered states of consciousness using their minds to gain access and eventually pass through a threshold or gateway into another independently existing reality. The shaman becomes an active participant of the other side, capable of developing and maintaining an on-going relationship with the spirits of that realm. By repeating the journey, the shaman becomes familiar with both the geography and the inhabitants of these different realms. From the shamanic perspective, all things are thought to be endowed with spirit: animals, plants and even minerals possess a spiritual essence. Travellers of alternative realms have the ability to communicate directly with the spiritual entities they seem to encounter.

Astral projection and the possibility that such a split in consciousness actually takes place can only constitute proof for the individual. However, individual experience is no measure of proof when we look for scientific verification. Astral projection has never been scrutinised scientifically. We may still find that it is no more than hallucinations taking place in the brain of the would-be astral traveller. We can similarly condemn shamanistic journeys as no more than a colourful rendering of the spirits, which are merely the life forces in all living things. The possibility of being able to tap into the deeper, more sensitive nature of living organisms does not necessarily constitute a different realm altogether. Even though there is much scepticism from those who have never explored altered states of consciousness, it is abundantly clear that those who are capable of such journeys do regard their experience as real.

Neurological Brain Function – That is all there is, Bud!

Some sceptics point out that the brain can mimic OBEs. Neurologist professor Olaf Blanke and his colleagues at Geneva University Hospital in Switzerland were able to bring about OBEs through electrical stimulation of the right angular gyrus in the temporal lobe of a 43 year old woman suffering from severe epileptic seizures (September 2002 issue of *Nature*). They claim they triggered OBEs in the female patient by stimulating her brain with electrodes. The doctors did not set out to achieve the effect, but almost accidentally stumbled upon the simulated experience whilst they were actually treating the women for epilepsy.

Their work has since been directed at explaining mysterious incidents of people claiming they were able to exit their physical bodies and watch the experience from an above or outside body location. They found that the stimulation of one spot repeatedly, in particular the angular gyrus in the right cortex, caused OBEs. By increasing the current amplitude, the person reports a perception of leaving the physical body. The patients experience the ability to see themselves from above and are able to see their own bodies lying in bed. They also have a feeling of lightness and floating about two metres above their beds. Professor Blanke and the doctors believe the angular gyrus has much to do with the duplicating of visual information and it also plays a role in the brain's touch and balance representation of the body. The overactive angular gyrus or dissociation of these elements may possibly be the cause of an OBE.

Michael Shermer, author of *Why People Believe Weird Things*, says if the brain mediates all experience, then paranormal phenomena are nothing more than neuronal events. He questions what really happens when people report OBEs and feelings of floating above themselves, and wonders whether these elusive creatures and mysterious phenomena are part of our real world or merely exist in our minds. He thinks the new information adds weight to the notion that they are actually nothing more than the product of the brain.

We are constantly trying to understand the paranormal and the supernatural in order to make it normal and natural. In fact, the paranormal or supernatural is possibly no more than normal and natural phenomena yet to be explained. It is the job of science to solve these puzzles and provide natural rather than supernatural explanations. The problem with the tests on the stimulated brain is that they do not consider whether the experience is actually real after all and not hallucinatory as suggested. If we consider that the astral body and its consciousness is still part of the physical nature, we can see how it is still possible for the astral persona to separate from the physical body if stimulated as suggested above. Many more tests are necessary to consider all variables, rather than coming to the first convenient conclusion that fits one narrow parameter set by natural science.

Clinical Tests – Perhaps there is more after all

New studies on NDEs are envisioned to establish irrefutable proof that some kind of mind or consciousness coexists outside the physical body. Medical researchers are looking for proof that NDEs take place and that some part of us, that is also conscious, is able to exit the body and observe life and its surroundings from an out of body vantage point. Early conclusions of scientific studies suggest that human consciousness can survive physical death and that mind, or at least some parts of it, and the brain are perhaps two different things. Two groundbreaking studies are forcing scientists to re-evaluate the current scientific model of the relationship between the mind and brain with far less agnosticism from some sceptics. The problem with most tests conducted on astral travelling (OBEs and NDEs), is that they do not include a possible third feature of a soul-mind. It is not difficult to imagine that the soul-mind is responsible for the conscious part of OBEs and NDEs.

On 15 December 2001, the highly respected international medical journal, *The Lancet*, published an article by Pim van Lommel, Ruud van Wees, Vincent Meyers and Ingrid Elfferich on a 13 year study of NDEs observed in ten different Dutch hospitals. The report called *Near-death experience in survivors of cardiac arrest: a prospective study in the Netherlands* gave a detailed account of how 12% of 344 cardiac patients resuscitated from clinical death accounted NDEs. Some of these patients experienced a sensation of being out of body, whilst others were seeing the all familiar light at the end of a tunnel. One of the theories by sceptics is that these events are mere hallucinations brought on by the loss of oxygen to the brain (anoxia) after the heart has stopped beating.

Cardiologist Pim van Lommel (MD), who headed the investigation that started in 1988, points out that the results indicate that occurrence of NDEs cannot be explained by medical factors alone. All patients were clinically dead after experiencing cardiac arrest with unconsciousness resulting from insufficient blood supply to the brain. When this happens, the electroencephalogram (EEG), which measures the brain's electrical activity, becomes flat and the patient can suffer irreparable damage to the brain or will die without cardiopulmonary resuscitation (CPR) within a 5-10 minute period. If we accept the theory that NDEs are caused by anoxia, then all patients in their study should have had similar NDEs. However, a mere 18% reported having had such. He explains that NDEs are perhaps caused by psychological states induced by the fear of death. Only a small percentage of the patients said they had been afraid directly before their cardiac arrest. Most patients said the cardiac arrest happened so quickly that they had no time for fear beforehand. NDEs are therefore more common in patients that had no time for fear before the actual trauma.

Of the 344 survivors of cardiac arrest who were studied for frequency, cause and NDEs, 18% had some memory from their period of unconsciousness, 12% had a significant experience called a "core" or "deep" NDE and 6% had a superficial NDE. Patients who have survived several CPR's in hospital have a significantly higher probability of having experienced a NDE. Some scientists were surprised that the NDE patients recalled their experience with the same degree of detail when interviewed again several years later. The determining factor for establishing an OBE experience scientifically lies in ascertaining whether the brain is dead at the time of the OBE. If the brain is dead, then according to the scientific model there is absolutely no chance of the mind being able to function or produce any thought whatsoever. According to Van Lommel, an American prospective study of 116 survivors and a British study of 63 survivors of cardiac arrest indicated 10% and 6,3% of patients respectively might have had a significant NDE experience. In the Dutch study, about 50% of the patients even reported awareness of being dead or had positive emotions.

The possibility of consciousness existing outside the brain, when the brain itself appears to be dead, is an especially important outcome generated from this research. Dr Van Lommel points out that there are several theories aimed at explaining NDE. Since electrical "stimulation" of some parts of the cortex can induce NDE-like experiences, it is suggested by some that neurophysiological processes must play some part. Experimenters with drugs have also reported NDE-like experiences after the use of ketamine, LSD or psychedelic mushrooms. However, these recollections are more fragmented and consist of random memories. They are also not as panoramic as is the case with NDE. Another opinion is that NDE is thought to be a changing state of consciousness. In this supposition the memories, identity, cognition and emotion function independently from the unconscious body through retained non-sensory perception. This to Van Lommel seems unlikely since consciousness during NDE is completely independent from the normal physiological body and waking consciousness.

According to these studies, the explanation must be that consciousness is experienced outside the body when the brain no longer functions. The brain is considered clinically dead due to a flat EEG. Furthermore, the same happens to blind people who have also been able to describe realistic and verifiable perception during OBEs at the time of this experience. For Van Lommel these events push medical ideas about the range of human consciousness and the mind-brain relation to new limits.

These studies showed that even though patients are considered clinically dead with a flat EEG, they were able to report consciousness with cognitive functions such as emotion, identity, memory from early childhood and perceptions from a position out of and above their "dead" body. Clinically dead in this context means a brain that has absolutely no electrical activity in the cortex and a total loss of brain stem function, which is evidenced by fixed, dilated pupils and absence of the gag reflex. This raises the question of how such coherent thought can occur when the brain is dead.

Van Lommel takes NDE and consciousness further by pointing out that there is still a lack of evidence that consciousness and memories are localized in the brain, therefore it is scientifically unverified. Van Lommel questions how a clear consciousness outside one's body could be experienced at the moment that the brain is actually clinical dead. These studies question traditional views about how thoughts, consciousness or memory is produced by groups of neurons in the brain. To Van Lommel it seems paradoxical that coherent awareness and logical thought processes are possible under cerebral impaired conditions like these and it raises puzzling questions about our understanding of consciousness and how it colligates to brain function.

University of Southampton researchers have also recently published a paper detailing their pioneering study of NDEs, which suggests some form of consciousness and that the mind may remain intact after the brain has ceased to function and the body is clinically dead. The team spent a year studying people who were revived after suffering a heart attack. The patients were all, for varying lengths of time, clinically dead with no pulse, no respiration and fixed, dilated pupils. Independent EEG studies have conclusively confirmed that all brain electrical activity and therefore all brain function had ceased at that time. When brought back to life, 11% of the Southampton patients who survived recalled emotions and visions during this supposedly period of clinical death. Seven of the 63 heart attack survivors were interviewed within a week of their cardiac arrest and reported some NDEs. Four patients complied with the strict Greyson criteria used to diagnose NDEs.

Dr Parnia, a clinical researcher from Southampton General Hospital in England and the author of *What Happens When We Die*, says we know very little about the subjective experience of dying and the nature of our minds during this process. He suggests the findings need far more investigation and if the results can be replicated it would imply that the mind may continue to exist after the death of the body. Dr Parnia has found there are broadly three scientific arguments for the causation of these accounts. The first are "brain-based", according to which patients hallucinate due to disturbed brain chemistry caused by drug treatment, a lack of oxygen or due to changes in carbon dioxide levels. A second is psychological and suggests that the OBEs and vivid encounters with tunnels, lights and deceased relatives are all constructed by the mind to ease the process of death. The third possible scenario is transcendental, which suggests that the experience is spiritual and the existence of the soul and therefore continued consciousness and life after death is possible.

Homo angelicansis

Arguably, the best example of consciousness beyond the brain comes from a cardiologist, Dr Michael Sabom, whose book *Light and Death* includes a detailed medical and scientific analysis of a phenomenal NDE. The case study is of a woman named Pam Reynolds who underwent a rare operation to remove a giant basilar artery aneurysm in her brain that threatened her life. The size and location of the aneurysm made standard neurosurgical techniques impossible. The operation, which is nicknamed "standstill", required Pam's body temperature to be lowered to 60 degrees, her heartbeat and breathing stopped, her brain waves flattened, and the blood drained from her head. Now that is awesome by any standards. What it means is that Pam was put to death. Under no circumstances should any person be able to experience any consciousness whatsoever during such an operation.

Three clinical tests were performed to determine whether Pam's brain was clinically dead during her rare standstill operation. The standard EEG measured Pam's brain-wave activity, auditory evoked potentials measured her brain-stem viability and there was a complete blockage of any blood flow to her brain. Pam's EEG was silent, her brain-stem response was completely absent and she had no blood flow to or through her brain. After removing the aneurysm, Pam was restored to life. During the operation, Pam experienced a NDE and was able to give remarkably detailed OBE observations of what happened during her surgery. She was even able to describe the unique surgical instruments and procedures used for the operation.

The amazing phenomenon does not stop here. All of Pam's vital signs were stopped and her skull was cut open. Pam reported that she felt herself "pop" outside her body and hover above the operating table. She was able to watch the doctors working on her lifeless body from her out-of-body position and observed the doctor using a saw looking like an electric toothbrush. Pam reported conversations by nurses in the operating room and remembered exactly what was happening during the operation. All this time every monitor used to register life signs showed Pam was indeed dead and had absolutely zero life whatsoever. Pam's case study is one of the strongest examples of veridical evidence in NDE research to date.

It is premature to say NDEs are proof that the soul exists in multi-dimensions or some spiritual destination. However, it does not exclude the possibility that some form of consciousness resembling our soul nature enables out-of-body phenomena to happen. To say we are souls temporarily housed in earthly bodies is still speculative, but it is a good explanation for what we experience as humans in terms of our souls. This, however, does not make it true. Even NDEs, when people see light at the end of a tunnel and their lives flash before them, may be all there is soon after physical death. It is highly likely that our ethereal, astral and ego bodies die together with our physical nature. This would mean that all astral cognition ceases to exist when the physical body dies.

Scientists probing the paranormal are hoping to set up a further experiment to determine with more accuracy whether the mind can step outside the body at the threshold of death. The study would involve interviewing OBE survivors of cardiac arrests. The researchers intend to place special objects and pictures around their cardiac units to see if survivors claiming to have had an OBE would know about the objects afterwards.

Whoopy-do-da-day!

Triunism postulates that it is not actually the mind of the brain, but a secondary "mind of the soul" that can separate from the body. That implies we have two minds and perhaps even two sets of consciousness – a rational mind function and a spiritual soul function. The brain and its thoughts operate in our physical dimension and a soul-mind operates and deals with spiritual matters in an alternative realm. Strangely enough, there are those who already imagine themselves to have such extraordinary abilities, enabling them to tap into both minds at will. The problems with such claims are that they come across as schizophrenic behaviour. However, the concept of operating in two minds has become familiar for many people who find themselves thinking with one mind whilst assessing their thoughts with a secondary mind and vice versa. The ability is likened to being conscious of your consciousness or being in two minds at the same time.

Astral projection and NDEs are not necessarily validation for the existence of a soul. Not only was Pam's consciousness able to exist outside her body, but it managed to do so whilst her body and brain were clinically dead. Pam's case has a strong resemblance to what we think of the soul. Even so, we have no idea whether such consciousness is sustainable indefinitely. If consciousness rides on the back of the soul and the soul on the back of the astral body, it is not a sure thing that we can sustain this condition without the physical body intact. For now, this is the cut-off point where our trail runs cold. Pam's case is probably the most credible validation of consciousness beyond the body and beyond death. Even so, it is not enough.

In NDEs and OBEs part of our nature is capable of experiencing consciousness at a distant location away from the brain. Practitioners of astral travelling can experience brain-mind consciousness and distant out-of-body astral consciousness simultaneously. During astral projection, the body and mind are still alive whilst a secondary mind is projected. Pam's body and brain were clinically dead when she had an OBE. This is a strong indication that our brain-minds and projected-minds are not one and the same.

OBEs and NDEs are indicative of triunism and differ intrinsically from trichotomy and dualism. The mobile part of our consciousness is called the astral or soul-mind. In triunism, the projectable soul-mind can also be an informant of the brain-mind. We consist of a physio-biological body, electrochemical brain-mind and spiritual soul-mind. I exclude Biblical Paul's trichotomous spirit as the third member of our nature in this context. To me, the use of spirit resembles no more than the attitude with which we conduct our lives. It has very little, if anything, to do with the topic discussed here.

The soul-mind dilemma keeps popping up. A triune nature of the soul poses serious problems. Neurosciences falling under the disciplines of biology, psychology and philosophy determine that all thought processes are the product of the physical brain. In the scientific model, it is impossible to have more than a dualistic nature of body and mind, let alone a triunistic one with an independently existing soul. This reverts to the kind of stuff ancient civilization's mythology and mysticism was built upon.

The two minds in triunism is not something that we could have explained in earlier scientific terms. The concept of triunism is not entirely new though. For thousands of years people have thought God communicates with our spirit through a Holy Spirit. Others have guides and/or angels who communicate through spiritual visitations. I am of the opinion that all these voices, visions or visitations are nothing more than our own soul-mind communicating with us. For the first time in the history of life on Earth, the soul has become able to transfer higher information to the mind. Not only have we become conscious of the soul, the soul is becoming conscious of its own existence and is able to pass such information to the brain. It is self-aware. The soul is getting a mind of its own.

Chapter 13

Cosmic Soul

Souls are not found in our observable universe. No matter how hard you look for a soul in our universe you will not find one. If a soul exists, it does not exist in the observable physical dimensions of space and time. The best we can do is point to a person and say that perhaps that person has a soul. How do we recognize something that is still so ill-defined and beyond the realm of our sensory perception and technology? Regardless of how much insight into and detail we have of our universe, we have simply never found the soul anywhere.

Rather than giving life or perhaps even consciousness, we see the soul as adding to the nature of life. The most prominent feature of human nature is our enlightened consciousness, which supersedes other life forms. We know that we exist and we know that we think. We also know a substantial amount about how we exist and how we think. Humans have a dual mind perception whereby we can simultaneously be conscious and conscious of our consciousness. This was not always the case. Homo sapiens dates back a mere 150 000 to 200 000 years depending on what model we use. Homo sapiens did not stop changing over that time. Evolution did not stop there; we are subjected to a constant process of change. If humans evolve on a physical level, does the soul also evolve on a spiritual or less physical level?

Let us look at how we came to believe there is a soul in the first place. The creationist view postulates that God created heaven and Earth in six days. This dogma has been overshadowed by discoveries showing the universe as we know it to be approximately 14-20 billion years old and life on Earth has existed for more than 3,5 billion years. All modern scientific discoveries show that life was not created some 6-7 000 years ago as the Bible and Christianity proclaim. These discoveries relegate the biblical creationist doctrines to mythological status. Not that we are able to successfully explain all the mysteries the universe yields, but we are able to say the creationist doctrine is not true in a literal sense.

Without a proper definition of the soul, we have to look for evidence of some form of life able to survive death and yet retain consciousness. The soul must be an entity that is perhaps able to function outside the confines of a host such as our physical body. We can therefore say with confidence that for the soul to exist, two ingredients are required: life and consciousness. We can also say with a measure of assurance that our search must take place in an alternative realm, beyond our common perception of ordinary space and time. This definition takes us on a fascinating journey already established by many groundbreaking scientific explorers over the last two centuries.

Life and soul – our still illusive soul

To start our search for the soul we look at three important phenomena: the actual moment our universe was born, the inception of life and the genesis of the Ego or "I" consciousness. As we investigate the science of creation and the unfolding of the universe, we try to determine where the inception of the soul might fit in. Putting religious doctrine aside, we turn to science.

Most cosmologists today are convinced the universe came into being about 14-20 billion years ago with an explosion of gigantic proportions, followed by an expansion and cooling process. The American astronomer Edwin Hubble found initial evidence that the universe is expanding in the 1920s when he observed clusters of galaxies moving apart from one another. The expansion was also predicted by the general theory of relativity proposed by Albert Einstein in the early

20th century. How this initial point of creation came about is still beyond our ability to identify, and without more evidence, it is still meaningless to explore it at this point in time.

Our theories on cosmic origin and evolution can explain how the universe and life came into existence and how it came to be as it is today. We have improved our ability to observe and study the inner and outer fabric of our environment with the use of technology such as telescopes, microscopes and particle oscillators. We are even able to peer into our own bodies with the help of X-rays, Computerized Axial Tomography (CAT scans), Magnetic Resonance Imagery (MRI) and Position Emission Tomography (PET). These are all technologies devised to overcome the limitations of our senses. With the help of mechanical instruments like these, we are able to "see" and understand to some extent where we are as well as whom and what we are. Assuming it would one day be possible, our technology is not evolved enough to observe the soul as yet.

According to the Big Bang theory, the primary physical forces and fundamental particles that constitute the building blocks of matter were produced in the first split second after the "big bang". From about one ten-thousandth of a second after the initial point of origination, our universe had a temperature of a thousand billion degrees Celsius and the density spread evenly everywhere. As the universe cooled and expanded the photons and particle matter lost energy to a point where they did not have enough energy to make them interchangeable. This settling down of our universe into stable matter is still observed with advanced telescopes today in the form of background radiation.

As strange as it may sound, the Big Bang theory estimates that the vital materials of the cosmos had already formed by the end of the first second, but stars and galaxies had another million years to wait before they were formed. A cooling process accompanied the universal expansion from its earliest or original state to a point 250 million years later where the universe cooled to a chilling 100 degrees below freezing point. With these lower temperatures, possibilities of life emerged.

This point marks the critical level at which the density of ordinary matter becomes greater than the mass density of radiant energy.

The transposition from dominance of radiation to the supremacy of matter changed matter's behaviour and with the help of gravity, matter became more dominant. This crucial date and event determined the freedom of matter from radiant energy. Gas spread evenly and formed massive clouds, which eventually drifted apart as the universe continued to expand. This same gas formed our stars and galaxies. For another billion years, our galaxies and planets took shape during which water, ammonia, nitrogen, oxygen, and a few other ingredients that constitute life emerged. Several billion years later, we have our current universe with trillions of stars and billions of spiralling galaxies and planets including Earth, which was formed about five billion years ago.

The universe measures approximately 30 billion light-years across. That means it would take light travelling at the speed of 300 000 km/sec. (187 000 miles/sec.) 30 billion light years to cover the distance from one side of the universe to the other. A light-year is about 9,6 trillion kilometres, the distance light travels in a year. This is huge! Beyond this, we cannot see and have no idea what may exist. Some theoretical scientists are already speculating that there may be many other universes forming a multiverse.

Distances in space are massive and we do not live long enough to make space travel on an intergalactic scale possible. Our physics does not even allow for inter-solar space travel yet, let alone intergalactic migration. The biggest obstacle is not so much our physics, but our physio-chemical bodies. Even if we could live long enough, our bodies are unable to withstand the harsh conditions of outer space to allow space travel over such long distances. Our bodies only last about a hundred years at most and travelling constantly for this period will not even get us to the nearest neighbouring sun. Either we have to find faster ways to travel through space, or we must live longer. Then there is the mass factor: the same mass from the rest of our universe that forms part of matter originating from the big bang. Our bodies have mass and mass is difficult to project into and through space. If we had to find a way to get rid of mass, we may be onto something. Before we go there, let us first see how we obtained these strange living material frames in the first place.

Until the advent of life, there is no purpose to look for soul. We have no knowledge of the existence of life before 3,8 billion years ago when life formed for the first time on Earth. We have no proof of life other than our own. There are perhaps signs for the potentiality of primitive life on our neighbouring planet Mars, but we have no proof of any life but our own on Earth. Mars has relatively high levels of methane in what remains of its atmosphere that would be explainable either by reaction between basalt rocks and water or by methane-emitting organisms that inhabit watery areas. However, even if such life exists it would be too primitive to consider and I will therefore limit my search for soul to life on Earth.

Darwin and Evolution

Charles Robert Darwin (1809-1882) is a British scientist whose work has had massive repercussions for our modern perception of life on Earth. Although not the actual founder of evolutionary thought, he is the master of the modern evolutionary theory that claims all forms of life go through a slow-working procedure called natural selection. Darwin was the first to arrange enough scientific evidence in logical order to illustrate how all biological species change over time by natural means, in fact showing us it is one of the basic principles of life.

Darwin's theory profoundly introduced the idea that man was once part of the primate family. By implication, this means our ancestors are apes and not Adam and Eve as proclaimed by the religious texts of the Bible. Darwin's theory of common descent shows that all living organisms are derived from common ancestors and that all living beings ultimately have one common ancestor in the form of a single solitary bacterium estimated to have existed some 3,8 billion years ago. Micro-organisms on sedimentary rocks were discovered in fossils in Western Australia dating as far back as 3,465 billion years ago. Geologists working on ancient South African rocks have also found fossilised micro-organisms taken from 3,8 billion year old sedimentary rock at Barberton dating it more than 3,4 billion years ago. These micro-organisms used the sun to get their energy and they were able to make oxygen, which they released into the atmosphere.

First Biological Life

It is difficult to imagine that it was only two centuries ago that the human species became aware of evolution and its own animal nature and ancestry. Whether bacteria or primate, all of us live in one very tiny biosphere on the Earth's thin zone of air, soil and water. This is the only area in the entire universe we know of that currently supports life. It reaches a mere 10 km (6 miles) around the surface of the Earth. Phenomenal! Water is the most important component of the biosphere in producing and sustaining primitive forms of life, but the atmosphere and many compounds of carbon are also essential as the building blocks that make up the biological form of any living organism. There may be other fully functional biospheres or life-supporting regions in our universe, but we have yet to find them. Any biosphere must remain sufficiently stable for billions of years for life to form and to the point where it achieves some form of consciousness.

Creation of life on Earth either emerged from a single cell or from a group of organisms from the same species of cells from which all life on Earth descended. One of the first complete accounts of living matter originating from lifeless matter and its ability to perform continuous evolutionary processes was introduced to us in the mid 20th century by the Russian biochemist Alexander Oparin. Oparin called it "molecular evolution", but it is currently referred to as "pre-biotic evolution". Obviously strongly influenced by Darwin's evolutionary theory, Oparin explained how life is possible in terms of chemical and physical processes. Although the idea of pre-biotic evolution is already widely accepted, some aspects are still speculative. Most biologists and biochemists do not doubt that life originated on Earth itself and that it was the result of a sequence of chemical events. We do, however, recognise that the abundance in living forms of plant and animal life on Earth sprang from a single unique ancestor. Progressive evolution became much more complex with time and left us distant cousins of all other animals, plants and bacteria.

What is Life?

Bearing in mind that we are still looking for the first requirement (life) for soul to exist, we look at what actually constitutes life. Life is a term that can apply to the activities of all organisms such as plants, animals, bacteria and in some way even viruses. Primitive forms of life or the simplest of living systems consist of the basic structure of only one cell. All life in living organisms displays the characteristics of reproduction and evolution. The first unresolved and mysterious action of life is to reproduce by making identical or near identical copies of itself. I say mysterious since we are unable to explain the incentive will to recreate. The second characteristic of life is to change and increase the complexity of its form and nature through slow modification called evolution. The third requirement is that there must be an organic structure with enough energy for these processes to take place.

Bacteria are separated from their environment by a thin semi-permeable membrane, whilst plants and animals are multi-cellular. All living organisms use metabolic processes to produce a constant flow of matter and energy in their cellular network to repair and replace their structure. The organism is never in equilibrium and when new structures seemingly spontaneously emerge, it is called evolution. The genetic instructions for building a cell or multi-cellular organisms are written in a chemical code or genome and this is always composed of nucleic acid. It is usually RNA (ribonucleic acid), which is the messenger carrying genetic information. DNA (deoxyribonucleic acid) stores the genetic information and is responsible for the cell's self-replication. The DNA is the memory of the species and is copied and transmitted to each generation. The molecular structure of DNA is a double helix wound together. Reproduction of the genome takes place when the DNA's double helix separates and a new strand is created alongside each of the original strands, resulting in two identical DNA molecules.

Even though DNA is material in nature and made up of atoms like all physical objects, it is said to actually carry the message of life and form, as if from supernatural or divine design. The message of the simplest free-living bacteria is written in the genome of about 2 000 genes, whereas a human DNA requires billions of atoms.

What Informs Life?

Even though we can explain what life is in incredible detail, right from the very first bacteria formed, we still cannot say what caused that initial life form to make its appearance. The second unexplained and therefore supernatural event is most likely the formation of life on Earth with the forming of the first cell. Was there a God's creative hand at work? Evolution postulates that all present life descended from a single clone. The origin of life is still a mystery and scientists are still baffled by what made the very first microbe come into existence: chance or divine intervention. Since science does not allow for mystical speculation, it had to choose chance.

Paul Davies in *The Origin of Life* says that nobody actually wrote the message and nobody invented the code – life came into existence spontaneously. He rightly points out that some people find the idea of an instruction manual that writes itself by accumulating chance errors is hard to swallow. According to Davis, the designer was Mother Nature herself, working within the capacity of the laws of nature which are able to capitalise on chance. This property of autonomy or self-determination is difficult to determine. Nobody knows what physical properties of living organisms confer autonomy upon them.

Scientists will not speculate on what makes DNA tick. Is there a mind behind a DNA string able to replicate itself and evolve? What instructs the atoms in most basic forms of life such as viruses and bacteria to form and perform in the way they do? Occultism ventures into the realm of the unknown and offers answers to these questions, albeit answers we are unable to verify. I have shown in some detail that cognitive science and the scientific model discard the notion that soul marries mind and brain.

The question natural science and philosophy asks is whether the mind of a human being is physical or immaterial. Is the mind the product of the physical brain or does the mind require a soul to make consciousness possible? This is not quite the same as the question about the origin of life. In such context, we ask whether all living organisms require something like a life-giving force or a soul. Is a life-giving force needed to unite life and body? Again, scientists say there is absolutely no life-giving force behind any organism that displays autonomous behaviour or has the ability to replicate and evolve. The entire life process is completely spontaneous and self-generated. Any explanation as to how this life came into existence and how it is sustained is illusive and cannot be satisfactorily explained.

According to Davies the mystics have it wrong. No properly conducted scientific experiment has ever demonstrated a life force at work and we do not need such a force to explain what goes on inside biological organisms. There is no extra ingredient for its intended function. Likewise, in seeking to understand the origin of life, scientists look to normal molecular processes to explain what happened, and not to an external life force to enliven dead matter.

Is there some mysterious consciousness that informs life, divine or otherwise? In religious contexts, God is the giver of all life. In the occult, there are various elements that inform matter to the point where they become living organisms. Let us see what answers the occult sciences offer us. Even though the occult principles are rejected in scientific models, it is worth looking at. Occultists hold there is an invisible (to ordinary sight) intermediary realm between the divine creator and the physical realm that has life-forming attributes, which inform life in our physical realm.

Quintessence is one example of an ancient belief in a forming and binding force that gives structure to our environment. But what is quintessence? Empedocles, a Greek philosopher from the 4th century BC, stated that all things comprise the four primal elements of earth, air, fire and water. Aristotle also sketched the universe as spherical and finite made up of four elements. In occultism, there are usually five cosmic elements. The fifth is an illusive component called quintessence that is considered to be the binding element of the other four (some systems have seven, including Buddhi and Atma).

Quintessence in this context seems to have a vital and creative nature. Quintessence has many different names and is often called the Aether, Akasha, Akasa, Akasia or Akashya. Aether is a Greek name for the divine aluminiferous substance that pervades the whole universe. Blavatsky described it as a cosmically radiant, cool, diathermias plastic matter that is creative in its physical nature. In theosophy, it relates to a luminous fifth element which is invisible to ordinary sight and is known as the binding element of the other four elements. In early occultism, it was also known as the "Luminous Waters" or even the "Mercurial Waters". Modern occultism calls it the "Cosmic Memory". In occult lore quintessence is merely the fifth element of seven of which the sixth and seventh are even more illusive than quintessence itself. It is widely recognized that quintessence is already visible to clairvoyants and will in the future become visible to most all human beings.

Apart from the five elements of which the fifth, quintessence, is the binding element of earth, air, water and fire, there is the etheric life-giving force that motivates all life forms and organisms. Etheric forces lend form to life within our cosmos. The etheric level is said to exist between the physical and the psychic world. Etheric forces are the formative, integrative forces that work on inert matter to produce the rich diversity of natural phenomena, the growth of plants, animals and human beings, the cycles of nature, and much of the planetary phenomena observed from the Earth. All living forms are imbued with etheric forces, and indeed the phenomena of the kind and quality of life depend almost entirely upon the nature of the etheric forces at play within a given being. When the etheric forces are withdrawn, the body becomes inert and quickly disintegrates into its constituent elements.

Occultism teaches that the etheric forces keep form from disintegration, what we commonly know as death. There is also an etheric double or "life body" described by esotericism as a delicate body, invisible to ordinary vision, which covers the physical body and suspends it in a state of life. In almost every respect, this human etheric double is a slightly larger copy of the physical body. Under normal conditions the etheric double does not leave the body until death, but in certain individuals the link between the etheric and the physical is looser than normal, a condition which enables clairvoyant ability.

Informing this etheric nature is the life-giving force (astral) which exists at a spiritual level or higher level than the material realm. The astral force creates a replica or a conceptual image of the thought or idea, which can be transmitted into the physical dimension and then manifest as material objects. Something must have the ability to inject life into matter to create a living organism. Occultism infers there is such a force or conscious being that gives life, with much the same attributes as God. (For further esoteric insights into the mind behind the matter, see Chapter 19.)

Scientific definitions of life determine that organisms at the most fundamental level do not require spiritual forces such as etheric, astral and/or soul to inform them. Occultism holds that etheric is the minimum requirement all life organisms need at the basic or entry level. The spiritual counterpart is astral, and its memory is astral light, which is the equivalent of quintessence for the physical body. My, oh my, this is complicated! Presuming there is a soul, it means there is a higher playground for our souls or spirits before we reach the divine or Godly realm. If this realm exists, it is still illusive and undetectable by modern technology.

Scientists are renewing their interest in this area and are examining our accelerating expanding universe. Our universe is expanding at an even increasing accelerated rate. Where originally we thought the expansion was slowing down, modern physicists agree there are multi-dimensional realities and parallel universes yet to be discovered. Measurement of the data of supernova and cosmic microwave background data shows that the universe is flat and will expand forever. How this happens is a mystery, because initially it seemed there was not enough substance to make a flat universe. What's more, this picture of the universe implies that unseen components of "dark matter" and "dark energy" exist and are largely responsible for the way the universe evolved. Although no one has seen this illusive matter or been able to examine it and although no one knows what it might consist of, cosmologists are convinced that it shapes the universe. In addition, it is more prevalent than ordinary matter. Calculations and early evidence suggest that more than two-thirds of the universe is invisible or has simply eluded us due to our inability to observe everything that makes up our reality.

Cosmologists propose that the universe's acceleration has to do with quintessence. The cosmological dark matter and energy bear some apparent resemblance to the historical version of quintessence. Various ongoing experiments will provide the key to the quintessence hypothesis and give new insight into the nature and future of the universe. Physicists must understand quintessence before they can create a unified theory of all the fundamental forces. Dark energy was still a mathematical absurdity at the end of the twentieth century. Within ten years, dark matter and dark energy have rewritten our understanding of the universe. However, what is this dark matter and energy or quintessence according to science? In the 1930s quantum physicists discovered, much to their surprise, that no vacuum is ever truly empty. It is buzzing with activity, filled with particles and energy. Empty space is actually an incredibly complex substance. Physicists find the need to understand dark matter and dark energy more pressing than ever before, as they believe that this dark essence also found in the vacuum exists everywhere in the universe and may be forcing it apart. (See Chapter 18)

Surprisingly this is also what mystics have been saying for thousands of years: that empty space is everything but empty. With clairvoyant vision, you are constantly bombarded with animated visual textures and awesome spiritual forces in what is ordinarily thought to be empty space. Quintessence as a formative force, which has been claimed to be visible to clairvoyant occultists, may be true after all.

For almost three billion years, Earth was the dwelling of only a single kind of life – bacteria – that developed and multiplied everywhere where was water. The sea provides us with numerous fossil relics called stromatolites. These are stratified rocks formed by pockets of bacterial colonies using solar light. Marine animals and aquatic plant life emerged only 600 million years ago. Land plants began to occupy the continents only 420 million years ago. The first dinosaurs evolved and lived in the equatorial rain forests. However, a number of major catastrophes have caused their extinction over the last 600 million years. The first major extinction took place 440 million years ago and the last of the great extinctions – the one that caused all dinosaurs to disappear – happened 65 million years ago. Because determining data is still missing about a number of puzzling factors, the exact cause of earlier extinctions remains controversial. The cause of the last major extinction 65 million years ago, however, is well clarified due to a number of affirmative discoveries.

Somehow, in all of this we have yet to find a place we can call our spiritual home. In the scientific context with a dash of occult science, we need a home or dwelling place for our spiritual nature we call soul.

What would constitute a playground for our yet illusive soul nature? Nevertheless, before we try to establish an environment for our spiritual nature, the soul, let us see if we have found a satisfactory definition for our physical nature and its environment.

Chapter 14

Conscious Life

That evolution takes place is no longer questioned. Evolution shows that humans or Homo sapiens share a common ancestor. We can trace that ancestor back to the primate family still in existence today. Arguably, the oldest hominid remains are the seven million year old cranium and pieces of a jawbone found at Toumai in France (2002).

Chimpanzees from the primate family are the closest relatives humans have. Researchers have found that certain enzymes and proteins taken from humans and chimpanzees are virtually identical. Asymmetries in the left and right half of the brain of chimpanzees also resemble those of modern humans. Chimpanzees and humans share more than 98% of their DNA and 99% of their genes, which shows just how closely related we are. In fact, chimpanzees are more closely related to humans than they are to their nearest primate cousin, the orang-utans. Remember also that no two humans have exactly the same genes.

The earliest human-like apes were the Australopithecines. They lived in Africa from around five million years ago and started to walk in an upright position. They represented at least six species, all of which had relatively small brains and ape-like proportions. These primates probably developed bipedal locomotion to adapt to new habitats, although they remained tree dwellers and usually slept in tree nests like other apes. Then, about 2,5 million years ago, the Earth's climate was transformed by a change in the axis. As the planet spun away from the sun it caused an overall cooling and made climates more seasonal. Temperatures decreased and large areas of woodland were destroyed, leaving belts of savannah in their place. Some animals became extinct and others evolved.

The Australopithecines can be divided into two groups: the first containing the heavier-built forms of Aethiopicus, Robustus and Boisei who existed about 2,5 to 1 million years ago and ate rough vegetable matter. The second group was the lighter-built forms of Afarensis, Garhi and Africanus who existed about 3,7 to 2 million years ago and had a more general diet. Some scientists believe the Australopithecus Africanus evolved into the genus Homo: "The family of man".

Fossil remains show the oldest Homo species yet discovered to be Homo habilis, who lived about two million years ago and are thought to be the first human-like beings to use stone tools. Some believe Homo erectus may have survived in parts of Asia long after the species died out elsewhere. Homo erectus was more robust than Homo habilis and had a larger brain. There is, however, some controversy as to whether all the fossils of Homo at this time can be attributed to this single species. It is thought that the Homo erectus population spread from Africa to Asia presumably some time between 1,9 and 1,7 million years ago and it was the first hominid to spread out of Africa. Fossils assigned to this species have been found from eastern Asia and Java to Georgia in America. The most recent Homo erectus fossils from Africa, dating from about a million years ago, show an evolution towards Homo sapiens.

The shift to bipedal locomotion may not be all that difficult for a primate, in fact, the change from chimpanzee to Australopithecus is considered smaller than the evolutionary step from Australopithecus to Homo. Although considered the most important step in the history of humans, I would suggest that it is superseded by the next step when humans developed the ego consciousness (as described in an earlier chapter). Bipedalism was considered a very important characteristic as this upright posture freed arms and hands for other roles, in particular for the making and using of tools.

Homo angelicansis

Ernst Mayr in *What Evolution Is* describes the shift from Australopithecine to Homo as the most fundamental one in all of hominid history. This change resulted in the evolution of a series of important diagnostic characters: brain size more than doubled, teeth became much smaller, the arms shortened, legs lengthened and dimorphism, the physical difference between males and females of the same species, declined from a 50% to a 15% higher weight in males. The Australopithecines were threatened by other animals and had no weapons such as horns or canines to fend off these dangers. Some Australopithecine populations survived by using their wits to invent successful defence mechanisms like the use of fire and tools. They were the first humans to make flaked stone tools, and it is possible they used sharper flakes to construct lances. It is assumed they used fires to protect them. The point is that these descendants of the Australopithecines, who evolved into Homo, survived and eventually prospered. The integrated patterns of behaviour required for planning and fashioning tools were accomplished at least 2,5 million years ago. Some form of advanced code for vocal communication may have existed at this time. Archaeological records show the first recognizable tools made by humans were produced in East Africa about 2,5 million years ago. These discoveries also show how these humans had a unique capability for culture, social relationships and a sense of self-awareness.

A common misperception about the evolution of modern humans is that we evolved as a sort of relay race, where only one human-like species existed at a time developing from one form into the next. This is not true. Many hominid species evolved alongside each other for quite some time, but only Homo sapiens or human beings as we know ourselves today, survived. Scientists consider all living people today as members of this single species. Homo sapiens first began to appear 150 000 to 200 000 years ago and were equipped with technologies very similar those of the early Neanderthals, who were possibly the third last hominid species to exist alongside Homo sapiens. New finds indicate there was a hominid called Homo floresiensis or "Man of Flores" who lived in isolation on the remote eastern Indonesian island of Flores as recently as 13 000 years ago. Homo floresiensis are also called hobbits due to their small stature. They rarely measured more than a meter in their upright posture. (First published in *Nature,* 27 October 2004. Information from *Scientific American,* February 2005.)

There has been a long-running argument between scientists about when modern humans emerged from Africa. Some adopt the hypothesis that modern humans already emerged from Africa two million years ago and then spread throughout the world, and others believe our ancestors emerged only 140 000 years ago. The hypothesis known as the "Centre of Origin" postulates that a later group of Homo sapiens called Cro-Magnons spread from Africa to the rest of the world about 50 000 years ago. About 35 000 years ago, a wave of modern Homo sapiens reached Western Europe, and after several thousand years of coexistence, the Neanderthal disappeared. The extinction of the Neanderthals seems to be a direct result of the migration of Cro-Magnons to Europe.

Remains from Homo neanderthalensis date back to about 150-200 000 years ago and have been found in Europe and South-West Asia. Neanderthals had a large brain, heavy brow ridges and a muscular build. The exact cause or causes of their disappearance are still under debate. No remains of Neanderthals have been found later than 28 000 years ago and genetic studies indicate Neanderthals did not evolve into modern humans, but became extinct. Homo sapiens did not continue where Neanderthals died out, but were their contemporaries.

The question whether Homo sapiens and Neanderthals interbred to become part of today's Homo sapiens lineage has not been answered. Skeletal remains of a child with robust body proportions found at the Abrigo do Lagar Velho rock-shelter in Portugal's Lapedo Valley suggest that these two species may have interbred. The skeleton resembles that of a modern human overall, but has some Neanderthal traits. There is no further evidence suggesting that Neanderthals survived in any way. Modern humans generally have skeletons that are more delicate and DNA comparisons show that Cro-Magnons differed widely from the Neanderthals, possibly too much for them to crossbreed.

Three areas of human uniqueness are identified as the ability to use fire, the use of language and the size of the brain. Evidence of fire use was found in Kenya dating back 1,5 million years ago where soil sediments appear to have been exposed to high temperatures. Admittedly, the burned soil could have resulted naturally from lightning and there is no evidence of cooking or of repeated fire use at this early time. Evidence of ancient humans controlling fire was also found in Israel at the Hula Valley site, dating back as far as 800 000 years ago. Evidence that is more convincing was found 400 000 years ago in Europe at the French site Terra Amata, where what could be simple fire hearths were found along with charred animal bones, presumably from cooking. Unfortunately, this is not sufficient evidence to say conclusively that controlled fire existed. Going back 100 000 years ago, we have abundant evidence of regular fire use at Neanderthal and Homo sapiens sites. This ability to make fire would have given Homo sapiens distinct advantages over other earlier peoples in providing protection from cold weather and keeping predators away from camp at night. Cooking would have provided digestible food and parasite free meat. In addition, it may have provided a platform for social activity.

Upright bodies and bipedal locomotion preceded the evolution of the large human brain. The most rapid increase in brain size in the history of man's ancestors took place when Homo originated. When the Australopithecine's environment was seriously threatened, an increase in brain size developed. From there brain size continued to increase in a punctuated evolutionary pattern. This is documented in the first Homo fossils. There was a period of relative inactivity beginning around 1,8 million years ago. Homo rudolfensis and Homo erectus were the earliest recorded species at this level of hominization. Homo rudolfensis had a brain size of about 700-900 cc. This is almost double the Australopithecus brain, which only averaged 450 cc. Growth in brain size slowed down after Homo rudolfensis only to show a massive and rapid increase in Homo erectus about 600 000 years ago, when the brain size increased to about 800-1 000 cc. This tendency continued until very recently in evolutionary terms, some 100 000 years ago, when the growth of the human brain tapered off to its current average size of 1 350 cc. Neanderthal brain size reached 1 600 cc, but that was probably due to their robust and taller size. It is surprising that the human brain seems completely unchanged ever since we became Homo sapiens about 200 000 years ago.

From here on the greatest evolutionary contribution to survival was intelligence. Brain size and its capability is the one distinguishing factor that renders humans so different from other species. Physically we are matched and in some cases even hugely outmatched by just about any other animal on Earth. There was seemingly no need for further increase in brain size since our cultural advancement from hunter-gatherers to large city civilizations with agriculture.

Simultaneous changes took place in brain size and human physique when Australopithecines progressed from semi tree dwellers to terrestrial living. The faster development of the brain is most likely the result of human speech and communication and once accomplished, larger brain size was no longer a reproductive advantage in more complex social living. It is thought that Neanderthals may have been first to use language. About 40 000 to 30 000 years ago significant linguistic advancements took place when modern Homo sapiens emerged.

Palaeontological records show that an increase in the size of the birth canal was ostensibly incompatible with the new bipedal walking, which necessitates an upright posture. Growth of the brain had to be shifted to the post-natal period, which means infants had to be born prematurely. Providentially the mother's arms were no longer used as much for climbing trees and could now be used to care for their infants. At birth, human newborns are now 17 months premature. This new ingenious evolutionary change allowed the size of the birth canal to stay the same whilst allowing the brain to enlarge after birth. It did, however, have its drawbacks since it became necessary for extended parental care after birth. The brain of the human baby almost doubles in size in its first year.

Accompanying the enormous development of the brain is the fact that children now have to add to their genetic information by learning from their parents and social group until they are adequately equipped to learn from their environment as well. It is therefore no longer an instinctive and immediately capable being that is born, but a dependent and helpless being. Only in species with highly developed parental care can such information be handed down from generation to generation without it being carried in the genetic code. However, information that can be transferred with most animals is pretty limited, whereas humans have the ability to keep learning and developing like no other species. The transfer of cultural information in humans has become a major aspect of life. Out of this new evolutionary step emerged speech and possibly language, which enables highly sophisticated communication.

Even though the word language is used in connection with the transmission of information in animals, the nature of their communication hardly exceeds systems of giving and receiving signals. To be a fully-fledged language, a system of communication must contain syntax and grammar. Even chimpanzees show little if no ability to learn language and grammar after fifty years of attempts by psychologists. Chimpanzees so far lack the neural equipment to adopt syntax and cannot talk about the future or the past. With the manifestation of language, humans were able to develop oral customs much earlier than the invention of writing and printing. The development of speech through these evolutionary changes exerted huge pressure on the enlargement of the brain. The parts involved in processing information, storage and memory are particularly well developed in humans.

The enlarged brain has eventually led to the development of arts, culture, literature, mathematics and science as we know it today. It has also led to our unique ability to be cognisant of our consciousness, to become enlightened and portray conscience. Humans have even become able to activate such conduct at will. However, humans have mostly learnt how to become spiritual and recognise a soul. This perception is one of the most fundamentally unique features of any living creature we know of. Humans are so far ahead of any other species that if evolution had to run its natural course, it would take millions of years for any species to catch up to our cognition. This makes one think what it would take for our nearest relative, the chimp, to have a well-developed soul, recognise its soul and then act upon it as humans do.

Cro-Magnons – Ritualistic Art & Burials

Some archaeologists believe the earliest evidence representing modern thinking can be attributed to Homo heidelbergensis dating as far back as 400 000 years ago. This is more than 200 000 years before Homo sapiens evolved. In 2003, a single stone hand axe was found buried amongst the human remains at a site called Atapuerca. Some researchers working on the site think it may be evidence of the first funeral rite ever found. The remains of 32 individuals were dumped there in the space of about a year. However, the evidence is inconclusive, controversial and not accepted by the majority of archaeologists. At Schöningen in Germany, three wooden throwing spears were found also dating back to 400 000 years ago. Even though scattered evidence of modern tools is found pre-dating Homo sapiens, none of them show real signs of symbolic or ritualistic thinking.

Arguably one of the earliest forms of symbolic thinking comes from a cave in South Africa called Blombos. Snail shells from an estuary nearby were pierced with a bone awl and possibly used to string together for a necklace or bracelet. The levels within which these beads and an array of other advanced implements were found dates back as far as 75 000 years ago. Some of the beads, tools, red ochre usage and engravings show signs of premeditated behaviour. Similar finds corroborating early symbolic thinking dating this far back are found at sites in Botswana, Kenya, Ethiopia, Tanzania and Israel.

There is evidence that Neanderthals demonstrated some form of caring for the frail and sick and at some point started burying their dead in shallow graves. Yet again, not all paleoanthropologists agree with the view that the objects found in the graves were placed there in funerary rites. It may be that the Neanderthals were perhaps preparing the dead for some belief in an afterlife. It is, however, commonly accepted that the Cro-Magnons were the first to display routine and detailed burial that started to show signs of ritual and belief in an afterlife from about 40 000 years ago.

Ian Tattersall, curator of physical anthropology at the American Museum of Natural History in New York and author of *The Monkey in the Mirror*, explains that over 30 000 years ago the Cro-Magnons were creating astonishing art on the walls of caves. They had mastered the art of fashioning instruments from stone, bone and ivory. Clay was also modelled occasionally, which may have had magical and/or religious functions. They ornamented their bodies and buried their dead with elaborate grave artefacts, which Tattersall believes is probably an indication of their belief in an afterlife. Their technology included innovations such as nets, textiles, ropes and even ceramics. The Cro-Magnons were a species much like us and their relationship with the rest of the world was unprecedented in the entire history of life.

Tattersall explains how the lives of the Cro-Magnons were also drenched in symbolism. Nothing like them appears in the record left by any humans that came before. They were truly a new kind of being with a display of ritual most strikingly seen in the burial of two young individuals and a 60 year old male, who were buried with astonishing material richness at a site in Russia called Sungir. The practice of burial rites that include items placed in the grave, combined with ritual, indicates a belief in an afterlife. Tattersall explains we cannot say for sure that the artefacts were made specifically to be used in burial, but it is certain that the knowledge of inevitable death and spiritual awareness are closely linked in Cro-Magnons.

It is here that we find the most ancient irrefutable evidence for the existence of religious experience connected to death and burials. The effort put into the aesthetic production of graves suggests that decoration, elaboration and art were integral components of the lives and societies of the people who made them and were not random, but planned. Art was definitely not just occasional or incidental. These societies must have been running considerable economic surpluses to allow the disposal of objects that were so valuable in terms of the time taken to make them. These people clearly did not have to devote all their time to the basic business of making a living; they were efficient exploiters of their environment allowing leisure time for symbolic pursuits of this kind.

The fact that there is a considerable variety in the elaboration and detail of Cro-Magnon burial (the sheer opulence of Sungir is one exception, rather than the general rule), hints at a social stratification and division of labour in Cro-Magnon society. Some Cro-Magnons were buried with an extraordinary abundance of artefacts of various kinds; others were more simply interred. Given the erratic sampling of Cro-Magnon burials available to us, some of this variation may simply have been due to difference in affluence of different societies overall. There can be little doubt that part of it, at least, reflects a differing importance of individuals in society. Some of that difference in status may well have been inherited, for it is highly dubious that children would have had the opportunity to make any significant mark on their society through their own accomplishments. What's more, it is hardly probable that these young people had made their richly adorned vestments themselves. It is much more likely that the sheer diversity of material production in their society was the result of the specialization of individuals in different activities. Those who ground the mammoth-tusk beads of Sungir and who straightened out those mammoth-tusk spears, may well have received far less elaborate interments when their own turn came to be buried. Contrasting this site with other Ice Age burial localities also draws attention to the considerable local variation in mortuary practices that existed during Cro-Magnon times.

In some places, bodies were flexed in the grave and some were stretched out. Some graves were covered with rock slabs; others were not. The nature and abundance of grave artefacts varied from place to place. Such differences in burial customs must have reflected a wider cultural diversity. Nonetheless, the material aspect of Cro-Magnon life that speaks to us most directly as human beings lies in the evidence these people left behind of art and symbolic representation. What is obvious, is that here we have evidence of highly complex symbolic systems. Whatever the case may be, the Cro-Magnons have left dramatic evidence of the integral importance of symbols in their lives.

This typically modern human cultural diversity and the pervasiveness of symbolic behaviour itself, stand in dramatic contrast to the relative monotony of human evolution throughout the five million years that preceded it. For, prior to the Cro-Magnons, innovation was extremely sporadic at best as far as we can tell. As I have hinted, there are suggestions of early art and symbolism elsewhere than in Europe, and doubtless in the late Ice Age similar developments were taking place throughout the inhabited world. Nevertheless, the record left us by the Cro-Magnons is so far unparalleled.

Eventually humans abandoned their nomadic hunter-gatherer existence and began to settle in permanent villages, of which some became towns and, eventually, cities. People began to cultivate wheat, rye, peas, lentils and other plants, and to domesticate animals such as cattle, sheep, pigs and goats. They also made pottery vessels for cooking and storing food, fashioned sickles out of stone for harvesting crops, and used flat stones as querns for grinding grain into flour. Although this change is known as a revolution, it happened slowly over thousands of years. Because farming life involves tilling the soil, planting, watering, harvesting and caring for animals all the year round, it is more labour-intensive than hunting and gathering.

Ritualistic burials have not always been part of human behaviour, as we know it today. It is something that slowly evolved over a period and, in essence, we can say that it was a characteristic displayed exclusively by Homo sapiens over the last 30 000 years. That means it started becoming part of our behaviour after other Homo species had already died out. It could be that the late Neanderthals picked up some of their ritualistic burial behaviour from their peers the Cro-Magnons, who had infiltrated the area. Their behaviour was not necessarily a natural display we can attribute to a belief in some form of afterlife.

Humans did not start to bury their dead some 6 000 years ago after the Biblical Adam received a soul implanted by God. A gradual awakening in the consciousness of humans over a period of at least 30-40 000 years brought humans to a place where the soul is recognised and incorporated in ritualistic behaviour. The well-established eschatology humans have today is the culmination of an elaborate ritual to care for the dead and the soul to ensure safe passage to the next realm.

Natural selection through evolution has significant repercussions for our understanding of the soul. We no longer question whether evolution exists. It is a fact and we merely study its mechanisms and implications. Evolution discredits religions, occult lore and many philosophical –isms. For the soul to survive our future attention, we need a new workable model that includes evolution and science. If we cannot place the soul in this context, we probably do not have one.

Chapter 15

The Ego/I Consciousness

Three events are extremely difficult to explain without reverting to mysticism: The birth of our universe from nothing, the inception of life from matter and the first time living organisms acquired consciousness. All of these appear supernatural because of our inability to explain them. It is almost impossible to imagine these occurrences without the help of a preceding "God". Each of these events sparked a sequel of events that we are able to observe, analyse and marvel at today.

The manner in which our cosmic events unfold makes them appear choreographed by a "God". Life on Earth and in particular our human conduct similarly seems choreographed. Our conscious participation in the universal phylogenies shows the consequence of our conduct. Human participation affects the outcome of at least part of the evolutionary process. Our conduct sparks a sequel of events that tells a story which adds to the evolutionary story. The conscious decision-making that humans display makes it appear as if we can determine the outcome. The reason for this remarkable development is our amazing brainpower, intellect, self-awareness, advanced consciousness and technology.

Humans have consequently come to think we are able to maintain our consciousness without our physio-biological bodies. To facilitate consciousness without the body and the brain we require the soul as a vehicle. Seen against the backdrop of many unexplained "supernatural" phenomena, the soul is not all that supernatural at all. The soul may have been ill-defined until now, but we are certainly able to find credible answers to the purpose and workings of the soul.

In occultism, soul consciousness is part of the Ego or "I" consciousness. Ego/I consciousness, which is the knowledge of the ego personality, surpasses mere intelligence since it is aware of its own existence. It no longer needs to act on genetic instinct or sensory perception; ego consciousness acts on self-induced rational command. How this process works still eludes us. The development of Ego/I consciousness took place simultaneously with Homo sapiens' perception of spirituality and the assumption that our conscious constituent could survive death. From an esoteric viewpoint, the Ego/I consciousness makes humans identify with an independently existing soul. For those who accept spirituality, the Ego/I consciousness is not the same as the mind, which is created by the brain. Our brain-mind accompanied by our thoughts and memories are studied from a cognitive and neurobiological perspective. In occultism the Ego/I consciousness is different from the neurobiological mind. It derives from our astral persona.

The Ego/I consciousness was not always present in Hominids. It similarly appears that the soul was not always part of human nature either. It is only in the latter stage of Homo sapiens' development that we see how the Ego/I consciousness came to the fore. Almost the entire evolutionary process took place without any conscious awareness thereof by any living creature, including humans. We merely study the results of evolution in hindsight. Even though we have discovered the existence of evolution, most of its inner workings are still obscured. Yet, evolution still takes place. It did not stop when we became conscious of its existence.

All life forms on Earth including humans are continuously caught up in the process of evolution. For as long as living organisms procreate, evolution and change will take place. Until now, nobody has dictated the biological process of living organisms. All life forms were merely participants in the process of evolution. It is only in the last 200 years that humans became aware and therefore observers of evolution. The moment we became aware of the long and arduous process of our own making, our technology became so advanced that we were able to dictate the direction biological evolution must go. Homo sapiens is getting ready to become "god" of its own physiological destiny.

The problem with our capacity to change our nature through genetics is that we change the direction evolution was going. If we do that, best we know the exact outcome and consequences. Chances are that our vanity will determine future changes to our nature. Presently, the emphasis is on either aesthetics or longevity. Cosmetic benefits will comprise more attractive or physically stronger offspring than the neighbours. This does not make us a fitter species. The scientific benefits of genetics are far more important.

The most important long-term consideration for genetic engineering is to secure the survival of our species. Science has long been aware of the evolutionary benefits of genetic adaptations to our species. To facilitate long, arduous flights to unknown destinies in space, we have to re-engineer our biological nature. Long distance space flights are far too taxing on our current physical bodies. To overcome this problem, scientists contemplate genetic changes to create more suitable super humans, capable of living in the harsh environment of space. We can easily be blinded by our ability to change our nature for interstellar space travel. Humans by nature are extremely adventurous and will sometimes jump at new opportunities like space flight and longevity without investigating the full consequences of such actions.

To prescribe the biology of our own nature will not only change our entire physical nature, it will also change the very nature of evolution. The moment we start to interfere or intervene in the natural processes of evolution by dictating its outcome, it becomes necessary to understand the full extent of the process and consequence of our interference under all circumstances. Before we get too cocky with human genetic engineering, we should recheck our premises regarding the soul.

I suggest the next natural step in evolution is the subtle ontogeny of our soul nature. Even though the soul exists on a more subtle level, it is still part of human nature and must evolve. Instead of making another physical and outwardly obvious change, humans are evolving on a spiritual (less physio-biological) level. Like every other aspect of life and evolution, the change towards a soul-empowered human is a necessity in the struggle for survival. I suggest that evolution through natural selection is in the process of creating a new species, called Homo angelicansis. Rather than obvious physical changes like upright posture and an enlarged brain, the most obvious difference between Homo sapiens and Homo angelicansis is an advanced soul.

In previous chapters, I showed how the Ego or "I" and/or the soul is the reincarnating vehicle of consciousness dwelling within the sheath of physical, etheric and astral on the material plane. This is the determining occult criteria when humans are evolved enough to have a spiritual soul nature. The Ego/I, which is part of our soul nature, distinguishes humans from fellow creatures. If humans had a spiritual nature more than 30 000 years ago, it must have been instinctive and dormant. However, awareness of the soul is evident from about 30 000 years ago. The Ego/I consciousness became more obvious when our speech and language developed syntax, grammar and communication of complex concepts. Even early cave paintings and burials did not include a perception of the ego-self. Conscious egotism is a very recent feature of human nature.

The Old Testament text of the Bible attributes this very feature to God, as in Exodus 3:13-14: "*And Moses said unto God, Behold, when I come unto the children of Israel, and shall say unto them, The God of your fathers hath sent me unto you; and they shall say to me, What is his name? What shall I say unto them? And God said unto Moses, I AM THAT I AM: and he said, Thus shall thou say unto the children of Israel, I AM hath sent me unto you.*" (*Holy Bible – King James Version*)

Homo angelicansis

By the time ancient Mesopotamian civilizations formed with temples, planes, houses and even burial chambers, we already had a highly developed spiritual consciousness. It is not entirely clear how this consciousness developed since we do not have a written record of it. However, it is abundantly clear that religious eschatology was well established around 7 000 years ago. In esoteric circles, only Ego/I conscious beings with souls can pass onto the next life. It is possible that only a small sector of our species is going to complete the evolutionary journey to become fully-fledged soul-beings: Homo angelicansis. If science embarks on a pseudo genetically re-engineered biology for humans or perhaps a cyborgian journey, what are the consequences for natural selection in evolution? I hope that enough people with spiritual wisdom and capacity are left to allow evolution to take its course. Either that or else we have to invent astral genetics.

Evolution has another ace up its sleeve. Slowly but surely, as if by divine intervention or some unknown directing force, we are being shaped and formed. Even though humans have come a long way in understanding evolution, most is still an unconscious process. Evolution is mostly recognized and studied in hindsight. That was still acceptable before we were able to dictate evolutionary changes. However, we are now able to dictate future outcome. That is why it is crucial that we ask in time whether humans perhaps evolve spiritually as well. Evolution does not take place on a physical level only. Any large-scale humanly premeditated intervention of our physical genome will influence or perhaps even terminate this vital evolutionary step.

The most outstanding new feature any human can have is the soul. This has never been seen as an evolutionary step! If it is true, it would be astounding and irrefutable proof that Hominids have indeed evolved a new aspect. Christianity through the Bible assumes that Adam was the first man and the first to receive a soul from God. That cannot be, since there were other humans on Earth at the time. Seen from a purely anthropological viewpoint, we have many bones and artefacts as evidence of preceding Homo species existing way before.

The only assumption one can make in biblical context is that all other humans in Adam's time were soul-less beings.

We can have two different species living side by side again for the first time since the extinction of the Neanderthals from Europe and the Hobbits from Flores. In the same way that the Hobbits from Flores remained undiscovered due to a lack of evidence, we can also say that a new species like Homo angelicansis is still illusive due to a lack of evidence. There is another option, but it is a ghastly one to contemplate. Perhaps all humans have become spiritual with a newly formed soul nature, but some have become Homo angelicansis after the nature of good and other Homo daemonicansis after the nature of evil, illustrating a struggle between good and evil.

The name Homo angelicansis is indicative of an angelic being with specific qualities and abilities accredited to it. Most people are familiar with the characteristics of angelic beings. According to Fred Gettings in *Encyclopaedia of the Occult* angelical beings function as cosmic powers or as heavenly warriors. Customarily angels are the link between a creator and humans living on a three dimensional Earth. They are also accredited with the ability to bring messages to guide and protect us during the spiritual transformation that is taking place on our planet. Major religions believe angels have an important place in the overall plan of creation. In occult lore, angels belong to a kind of celestial hierarchy of Nine Orders. The incorporeal angelic spiritual beings belong to the Ninth Hierarchy, which is above the realm of man.

The abilities equated with angels are their ability to transcend space and time. Angels can enter and exit our physical realm and alternative realities at will. We also equate angels with the realm of God and a higher dimensional existence. In the Old Testament of the Bible, we see that life on Earth took a new turn when the human soul was born for the first time. In Genesis it says "and man became a living soul", the soul status being the distinguishing factor that set man apart from the rest of the animal kingdom, and that includes Homo sapiens. Humans could have been evolving to this point for millions of years as the evolutionary theory postulates. The soul was added afterwards, almost as if part of a new grand design.

Genesis contains one of the earliest and most significant scriptures with the very same hypothesis that the soul set some humans further apart from the rest of the animal kingdom and from members of their own kind. We have branded Homo habilis as the handy man/woman (he/she who exchanges two feet for two hands), Homo erectus as the upright man/woman (he/she who walks on two rather than four legs) and Homo sapiens as the brainy man/woman (he/she who knows that he/she thinks). We can now classify Homo angelicansis as the angelic man/woman (he/she who walks with the body of an angel).

Evolution is simply a change in the gene pool of a population over time. Genes as hereditary units can be passed on from one generation to the next. The instructions of our makeup are locked up in the sequence of our DNA. Through procreation and reproduction, mutations take place in the genetic material from one generation to the next. The change in the reproduction process caused over many generations is called evolution. In biological evolution, mutations occur over generations of organisms, and not in the lifetime of a single individual. Individual organisms cannot evolve, since they retain the same genetic material throughout their lifetime.

It is sometimes said that enough mutations in a species cause a new species to evolve. However, it must be added that if enough mutations take place in a specific sector of one species, a new and different species can result from that particular master species. Changes in a sector of a species is also evolution. It does not necessarily take the entire species to mutate together or in unison to have a new species evolve. In evolution, it is the fittest that survives. Those who have adapted to changes in the environment and keep up with future challenges are the ones most likely to survive. If the soul is the requirement for making a portion of our species the fittest for future survival, then that portion and its offspring with souls will survive with time.

If the entire species changes over time, it remains the same species but is more evolved. Even if the species changes dramatically and becomes different to what it started off as, it is still the same species. However, if mutations take place in a particular sector of a larger group of the same species, that allele strain can eventually become so different that it can result in a new species. Due to the slow process of evolution, the sub-species can evolve and exist alongside its master-species for some time without it being evident that the secondary species already exists. Initially the changes will be too small to recognize.

The theory of evolution postulates that organisms with the fittest characteristics necessary to survive and reproduce are the ones that flourish and survive. The ones that lack the necessary characteristics will leave fewer offspring or even die out. If humans interfere with the process to manipulate and alter the genetic code from one generation to the next, it is no longer a natural selection. Directive evolution through interference means that humans determine the evolutionary outcome of a species. For humans as well as any other species at the mercy of humans, it could be the end of the road for evolution by natural selection and survival of the fittest.

Humans have become so developed we are able to re-engineer any biological organism through genetic manipulation. Our intellect and technology can now override and redirect any evolutionary attempt of the natural process. Evolution is now in the hands of human "gods". We have come to the end of evolution through biological natural selection. Humans can now manipulate and genetically engineer any life form to anything we want – as long as it has biological nature. What was previously considered the creation of an "Almighty Creator God" can now be remodelled and changed in any way, as long as it is on a physio-biological level. This can be a phenomenal blessing, but can also be extremely dangerous. If we take over the role of "God" or the "natural process" so to speak, can we recreate the fittest substitute?

Until now, we were merely able to transplant individual organs of one body to the next. This in itself is revolutionary in the sense that the body of one organism is able to successfully carry and use an organ of another of its own kind. Prolonging life through organ transplant is similar to longevity achieved with genetic engineering. However, with genetic engineering the process of longevity can be on a massive scale and far more successful than ever imagined. Our ability to speed up change through technology supersedes the slow process of evolution. Nevertheless, in effect we are creating a new monster. By speeding up longevity through transplants, we are slowing the evolutionary process down since it means people can live longer. By speeding up change though genetic engineering, we can perhaps bring evolution to a halt in some respects. The main objective of genetic manipulation is to produce successors that are more advanced. Since we do not yet know everything about our nature and environment, we do not know what it takes to be "the fittest". All we are really attempting to do at present is creating new generations fit enough for our space migratory fantasies.

By prolonging and perhaps indefinitely extending the life of our physical bodies, afterlife becomes obsolete. Nevertheless, death is actually good for evolution. Offspring introduces new alleles with every new generation. Evolution can only take place through transferral of genes with mutations from one generation to another. With prolonged life, we can end up with a situation where we reintroduce elderly genes into the newer gene pools of later generations. That in itself can have a retrogressive effect on evolution. However, if we live to be a thousand years, we reintroduce elderly genes into newly evolved gene pools that have already adapted to new conditions. In such a case, we could substantially slow down the evolutionary process. The extreme case scenario is immortality, whereby we simply stop mutations from taking place altogether. Evolution through natural selection in such a case stops altogether.

Even if we achieve physio-biological immortality, there is still the chance of accidental destruction of the body leaving us with physical death. This means we still have to consider the soul as an alternative to physical immortality. Neurobiology and mind-science maintain the soul does not exist and all our spiritual experiences are hallucinatory, all in the mind. If this is so, we need no longer consider the soul as an alternative and find ways to reproduce and keep the body intact. However, if all human effort fails, Homo angelicansis is nature's great alternative.

If we look at evolution and its impact on life one cannot help thinking that some of these principles should apply to the soul as well. If the soul is the result of natural evolution of our species, then we can ask whether there are soul genes responsible for the existence and evolution of the soul. Anything that evolves must have a genetic code of instructions to allow the mutational changes required for evolution. The same would apply to the soul. The soul or its astral counterpart, as a spiritual extension of the physio-biological body, should exist on a level yet unobserved by our physical senses and current modern technology. Are these soul genes physical and visible or are they from alternative vibratory levels and/or alternative dimensions and therefore invisible as occult lore suggests? After all, even though we cannot see the etheric and astral bodies, they are still part of our physical nature. Their elusive vibratory and dimensional nature can conceal them from ordinary sight, but, they are "physical bodies" all the same.

The possibility of some kinds of genetics beings responsible for the makeup of spirituality has been emphasised by Dr Dean H Hamer in his book *The God Gene.* Hamer, a geneticist, takes a bold step towards the possibility that genes may be responsible for our spiritual experience. Hamer, who also discovered the controversial "gay gene" in the 1990s, argues that at least part of the level of spirituality humans experience is coded into our genes. He argues that the "God gene" is in the human genome known as VMAT2, which affects the brain's neurotransmitters. Two different versions of this gene exist. People with the God gene version show signs of being more self-transcendent. If this were true, it would mean that spirituality is merely a basic human inheritance and nothing more than an instinct. Hamer admits that many genes may be responsible for our spiritual experience rather than just one.

Even though the genes responsible for our spiritual experience described by Dr Hamer are not entirely the same as genes responsible for a soul, we can still question the validity of both. Spiritual genes responsible for the soul would make the soul an adaptive evolutionary trait. The etheric and astral bodies are commonly known in occultism as the soul carriers and seen in this light, we should have a genetic structure and code for the soul body to exist. In such a scenario, we would be able to say the soul evolves. Imagine if it has evolved on a less dense physical level that is still unobserved, some form of orphic-evolution where change still eludes our senses and technology.

In October 1990 The Human Genome Project (HGP) was started to gain a basic understanding of all the approximately 30 000 genes that make up the human blueprint. This international effort was coordinated by the US Department of Energy and the National Institutes of Health and carried out by numerous research companies. It cost $3 billion and took more than a decade to complete. It was aimed at making all findings accessible for further biological study. One of the project goals was to determine the complete sequence of the three billion DNA subunits by a process known as sequencing. At the turn of the 21st Century, roughly 90% of an organism's genes have still no determined function at all. Though the HGP is completed, analyses of the data will continue for many years.

Even though the genome provides an unparalleled understanding of the key composition of human genes and chromosomes, it can only identify the genes in the nucleus of a human cell and establish where those genes are located on the chromosomes in the nucleus. It cannot yet identify all the genes and their functions. Our process of understanding our genetic blueprint is limited to linkage mapping and physical mapping. Linkage mapping identifies the genetic "landmarks" order along a chromosome of which each marker has a unique DNA sequence. Physical mapping is a group of more precise methods placing genes or landmark sequences at specific distances from one another on a chromosome. The complete human genome sequence enabled us to understand and combat human disease with new cures. We also have a much greater understanding of human biology and genetic individuality. The human genome sequence enables us to understand our human physio-biological nature better, but cannot in its current form assist us with our spiritual nature.

My challenge is not to discredit the theory of evolution or human genetics, but rather to show that these processes do not only take place on a biological level in the observable physical realm as we know it, but permeates the so-called unseen spiritual realm as well. For this, we need far more advanced technology capable of measuring our less dense bodies in the etheric and astral realms or perhaps in hyperspace (see Part 4).

We have historically credited angels with the ability to operate in a supernatural realm out of reach of human perception. Human intelligence is far more developed than is human insight of the soul. Demands on our rational function and logical application of the brain far exceed our spiritual acquisitions. This is why our soul consciousness is lacking and trailing so far behind. It is all IQ – intelligence quotient rather than SQ – spiritual quotient or soul intelligence. Homo angelicansis is a species that relates to humans with high soul consciousness, and both the physical specimen and the spiritual counterpart are subjected to similar processes of evolution.

Soul Evolution

Evolution is divided into microevolution and macroevolution. Long-term changes fall under macroevolution. In the case of etheric and astral evolution (which may include soul evolution), it would similarly be part of a long-term change that stretches over many generations. Seen in the context of evolutionary principles the soul, like the physical body, cannot evolve in one lifetime. The soul of an individual cannot evolve, but the soul nature of humans can evolve over many generations. You cannot evolve your soul as an individual, you can only become aware of how evolved your soul is. For the soul to evolve one must therefore look over a period of generations to see how it collectively changes and adapts. It takes generations for anything to evolve, and that goes for the soul nature of Homo angelicansis as well.

Becoming aware of one's soul nature and spirituality is not evolution in itself, it is enlightenment. You become conscious of yourself and your evolved state. In this instance, some people are more spiritually enlightened and may be further evolved as well. Some people have more evolved souls than others do, and this has very little to do with IQ and a lot to do with SQ. You can become more religious or enlightened, but you cannot become more spiritually evolved. Your IQ can be off the measurement chart, but spiritually you may be lacking. Other than utilizing what you have to the best of your ability, there is simply nothing you can do about the state of your soul. Whatever you do, you cannot evolve in this one lifetime to be more than you are, you can only live up to your potential. Even then you would not be able to discover all there is to know about yourself. There simply is so much to know and too little time to become completely enlightened before we die. Our offspring and later generations will have the fortune of achieving and witnessing what we in our lifetime struggle to comprehend.

There are three schools of thought regarding when we first started becoming spiritual. The first is that the soul came about in one creationist act. A "God-given" soul does not evolve, because it is fully developed in every human and expresses itself simply by either doing good or evil to be judged after physio-biological death. In this case, salvation is mostly achieved through redemption and/or good works. The second hypothesis is that the soul comes from some alternative spiritual realm and invades human embryos in order to have an earthly experience. The soul is born anew many times in another body through reincarnation, which allows it to gain experience and become more enlightened. Then there is my theory that the soul is the product of evolution itself. Our soul nature is passed on from one generation to the next and with time will reach maturity for its future cause. With an evolutionary soul, life has managed to extend its existence beyond the physical realm. Our spiritual nature supersedes the physical realm and the phenomenon is becoming a conscious process.

The biggest problem we have with spiritual anthropology is the lack of fossil records or genes to study. Our technology is not developed enough to assist us in our exploration of this field of study. Not even Kirlian photography has been able to present much more than an image looking like a manifestation of the spiritual aura or "life force", which allegedly surrounds each living being. In 1939, Semyon Kirlian by accident discovered a process that uses pulsed high voltage frequencies to take pictures of usually invisible and radiating energy fields. The energy fields that surround us all, are considered by many today to be the paranormal world of auras.

We have to find alternative ways to define and study our soul evolution. The soul and its host, the astral body, are not obvious from our known dimension of space and time. To include an alternative dimensional existence to the current timeline of fossil and genetic evidence becomes tricky if not impossible. Certainly, the majority of humans experience having a soul and constantly try to define it. We even seek ways to preserve it during life on Earth and especially to maintain its existence during the afterlife. This kind of behaviour seems unique to humans, as no other species holds the preservation of the soul as part of their experience of conscious life on Earth.

Scientific data shows humans to be at the pinnacle of evolution with regards to intelligence, technology and consciousness. With further analyses, it appears we also have some form of multidimensional spiritual experiences beyond the mere physical daily existence. Our awareness of alternative realms brings about practices such as taking care of the dead and rituals to increase the chance of spiritual immortality. The idea of the soul coming from an alternative soul realm and entering the human at embryo stage seems quite inconceivable, but not impossible. However, with our new knowledge on archaeology and anthropology it seems more likely that humans evolved to be as spiritual as we are through natural selection. Either way the result is the same. We may have a soul which is part of human nature and it seemingly is on its way somewhere. Whether it was given to us by "God" in a creationist act, whether it originated from aliens or extraterrestrials, whether it comes from a soul realm to experience life on Earth, or whether we are evolving and growing our own souls as life-forms on Earth, the soul is on its way somewhere, and that destination needs investigation.

Neither the explanations of our soul nature in previous chapters nor our evolutionary timeline gives a successful account of when exactly humans acquired a soul. The problem with the evolutionary timeline is that it only considers the physical dimension of space and time, but the soul seems to exist in another realm. Even though we have become aware of it in our realm of existence, it is actually not part of this dimension. To place our soul on the evolutionary timeline we must add another dimension.

It is eminently clear that we cannot point to a definitive creative act when we received a soul from a "God", unless we accept the creationist story of a soul-filled Adam. It is also clear that there is no room for the extraterrestrial invasion from other planets or dimensions. However, we do have proof of a gradual evolutionary unfolding of life on Earth. Life in itself achieved all it did on its own, and if it acquired a soul, it did so by itself. To prolong the soul's existence after biological death, we seem to look towards an alternative realm other than the one we are accustomed to here on Earth. It is possible that our awareness of the sky and the possibility that we needed to go there one day started our evolutionary step towards a transmigrating soul. Since we could not achieve this goal indefinitely with our bodies intact, we postulated such a journey after death with the soul. The soul somehow gave us that status to become immortal spiritual travellers.

PART 4

The Science of Spirituality

Chapter 16

Heaven and Hell

Even though the disciplines of religion and occultism have managed to explicate their views on spirituality and the soul, they are unable to legitimize their doctrines convincingly. We cannot hold ideological systems of this nature as true due to a lack of credible theory and evidence. To complete our quest for spirituality we need validation that the soul and a suitable spiritual realm exist. With the introduction of cosmogony, evolution and neurobiology, preceding doctrine seems circumscribed. For the soul to exist, it must exist somewhere and somehow. "Projectable" consciousness without the help of the physical body means that something like a soul makes it possible. Since near-death experiences and out of body observations can take place from alien vantage points, we also question the extraneous nature of the spiritual experience.

For a spiritual realm to exist we should effectively be able to prove it by means of our applied sciences. We have to keep in mind if and when we do find such a realm, it will cease to be spiritual and hidden after all. The concept of a hidden netherworld is found in many religions. God, gods and other spiritual beings dwell in heaven and our spiritual counterpart, the soul, is destined to be incorporated into it after physio-biological death. Ancient civilizations speculated that heaven was a physical place. The original meaning is literally "the sky", which referred to outer space or the dark areas of the night sky where there are no stars.

Homo angelicansis

Catholicism and Christianity teach that the final state of our spiritual achievement will be the conjoining of one's spiritual body with a resurrected body, which will live in heaven. In the book of Revelations, John explains how good souls will be resurrected and reunited with new, perfected and glorified bodies before entering this most unusual destination (*Revelations 21*). His description of the so-called godly realm of heaven is presented in the form of a "Holy City" called "New Jerusalem", which is detached from our physical realm. John accompanied God on a little sight-seeing tour during which he was shown that heaven is made of gold and glass with twelve layers, each comprising a different precious stone. So much for a spiritual realm! Early belief systems did not perceive heaven as an immaterial spiritual realm. Modern science challenged the dogma of a physical heaven and it has now been replaced by the view that it could be some unknown hidden dimension.

The Roman Catholic Church advocates an intermediary destination called purgatory. Purgatory precedes heaven to help cleanse one's sins and assist persons who had repented but had not atoned for their sins. It is a place where those who have died in a state of grace undergo limited torment. According to the doctrine, all who enter purgatory will eventually reach Heaven. Purgatory is most likely nothing more than a fabrication to cover for the Biblical time lapse between death and heaven. Since not all Christians die at the same time, the doctrine teaches there must be a time of waiting called "limbo" from death to judgment day and heaven. Since both are unique occurrences, a person's soul must be in limbo from the time of death until thousands or even millions of years later when judgment day comes to pass.

It has now surfaced that not even the Vatican is entirely sure about the state of limbo. On Friday, 6 October 2006, in an article called *Vatican to review state of limbo*, BBC News reported that Pope Benedict wants to amend or perhaps even do away with the doctrine. The reason given for the dismissal or change is that the theory that originated during the Middle Ages, was merely a solution to the theological question over the souls of babies who died and were too young to have committed any sins of their own. The Vatican is worried about the many people in developing countries turning to Islam, which deliberates that the souls of stillborn babies go straight to paradise.

Nevertheless, purgatory is not scriptural. One of the first ever documents to mention purgatory is a letter from Nicholas of Saint Albans, which dates back to only 1176 AD. Even then it was only clarified by the Catholic Church in 1254 AD. None of the Bible scriptures applied to clarify such a mythical place actually mentions purgatory or even closely resembles the concept (*Dan 12:10, Zech 13:9, Mal 3:2-3, 2 Mac 12:39-46, Mt 5:26, Lk 12:58-59, 1 Cor 3:13-15*).

Charismatic Christian teachings about entry into heaven rely heavily on the teachings of Biblical Paul. According to Paul, confession that Jesus is Lord and faith that God raised him from the dead, are the only two requirements for salvation (*Romans 10:8-10*). However, if we go by the writings of James, Jesus' supposed half brother, the Bible is contradictory. According to James, man is collectively justified by faith and works, and not by faith alone (*James 2:24 & 26*). It is not enough to simply have faith and verbally express one's belief in Christ, one must also live a good and decent life. The Catholic Church spices it up some more, making entry into heaven dependent on Christians receiving God's grace through activities such as baptism, the Eucharist and confession. The perfect life with God in heaven is a state of supreme and definitive happiness.

For Protestantism, entry into heaven depends upon one's faith in Jesus. In Calvinism, entry into Heaven is predetermined by the grace of God alone and all Christians have in fact already been chosen to be saved from the beginning of time. Arminianism disagrees and teaches that faith in Christ through an act of free will is essential. According to Universalism all people will eventually be saved regardless of their religious belief. What else can we conjure up?

The eastern equivalent of purgatory is Pure Land. With the emphasis on reincarnation, a large section of Buddhism is based upon the Pure

Land Sutras or Western Paradise Sutras. This concept may have originated in India around the turn of the 1st century AD. It later spread to China, where it became popular as early as the 4th century AD. Pure Land is central to the doctrine inspired by Amitabha Buddha. Amitabha is said to be a monk who was instructed by the 81st Buddha, called Lokesvararaja. In Pure Land, there is no need for meditation or ritual. The only requirement for rebirth in Pure Land is verbal repetition of the phrase "Homage to the Buddha Amitabha". Amitabha promised a far more lavish Buddha Land than his predecessors for all those who called on him.

The doctrine of Pure Land spread to Japan and slowly grew in prominence in the 12th century, when Honen Shonin established the Pure Land Buddhism sect known as Jodo Shinshu or Shin, today Japan's largest religious sect. The monk Honen also preached perpetual elicitation of Amida (Japanese for Amitabha) through dance and chant. Shin became especially popular amongst the general population for its simplicity and simple form of veneration. Pure Land Buddhism is only a branch of mainstream Mahayana Buddhism. In Mahayana Buddhism, fortunate souls first reincarnate before their final attainment of Nirvana.

Nirvana in Buddhism is the equivalent of heaven in Western traditions. To reach Nirvana one must find union with the cosmos and all human desire must be extinguished. Perfecting the soul is achieved through successive transmigrations. Since Nirvana is seen to be virtually unattainable from earthly life, Pure Land has become the intermediary phase before reaching the ultimate destination. Nirvana is impossible to define in any form of human communication. It must be realized through individual experience. Nirvana is not altogether the same as heaven. It is neither a place nor a state and can therefore be reached without dying. Nirvana derived from the Sanskrit entails "extinguishing", and the study of the sources and development of the word implies an extinguishing or "blowing out" of a fire or candle flame. In Nirvana, the flames of desire, lust, hatred, greed, suffering and ignorance are extinguished through endless cycles of transmigration. Through sequential cycles of reincarnation, we are able to perfect the soul. In the Buddhist context, the person realizing Nirvana is compared to a fire extinguished when its fuel supply has run out.

The mythical belief of a good God residing in heaven is further misconstrued in the antonym of the "Devil". This evil creature resides in Hell, meaning "hole" or "hidden": For the Babylonians it was a place where no one can return from, located within the planetary spheres. In Greek and Roman mythology it was called Hades and Tartarus, located somewhere in the underworld. Entrance into heaven is conditional, causing those who do not meet the criteria to go to a place of punishment. Zoroaster at the turn of the sixth century shaped most of the sinister imagery of hell. His imagery of a dark underground place laden with fires and molten rock where unbelieving evil souls are tormented after death, still lingers on. The Devil or Satan is more popularly depicted as a red creature with horns, a pitchfork and a long tail with a diamond shaped end. Hmmm!

Heaven in occultism is known as Devachan. Several deaths occur in occultism before Devachan. The first death takes place when the physio-biological body dies. After that, there is a gradual break-up of the lower principles or quaternary of man in Kamaloka (occult version of purgatory). The second death occurs when the etheric body dies during Kamaloka. This is said to take up to three or four days after the first death of the physical body. The third death takes place years after Kamaloka when the astral body dies. This allows the higher ternary of man to be separated from the lower quaternary and freed for the final destination of Devachan. In Devachan, all previous incarnations are realized on the highest spiritual level. Devachan is thought to be the dwelling place of God and nine other celestial hierarchies or angelic realms. The celestial hierarchies are dominated by spiritual beings also capable of manifesting in the physical realm. These beings are imagined to range from the highest realm of God to the earthly world we find ourselves in.

Homo angelicansis

Dionysius of Areopagite is controversially acclaimed as the first to define the celestial hierarchies in the 1st century AD. It was only as late as 553 AD that the pseudo-Dionysian division was presented to the Catholic papacy at the second Council of Constantinople. Pope Gregory I presented a slightly different order that reversed the scale from nine to one, giving rise to several difficulties in the medieval period. Even though Catholic liturgy still prescribed these principles for public worship, not many people question spiritual continuation beyond the common concept of heaven, hell and perhaps a preceding condition of purgatory. However, the modern division with its nine celestial hierarchies and appropriate spiritual beings persist. The highest orders range from spirits of love, harmonics and will in charge of planetary systems and time. The middle order of wisdom, movement and form involves the metamorphosis of matter. The lowest order consists of spirits of personality, archangels and angels who dictate and rule the domain of Earth and humans alike.

What a load of rubbish! There is much speculation about a ghostlike heavenly paradise awaiting us when we die. The reasoning behind it must be the finality of death. Once the body dies, the mind and personality dies with it. The soul, if it exists, simply disappears too, never to be part of this dimensional existence again. All that lingers on for some time after death is the "spirit" of the person in our memories. We do not find souls in our observable universe; if they survive, it must be somewhere else. According to religious doctrines, our soul somehow eventually makes its way to at least one of these unlikely alternative destinies, depending on the type of life led and/or salvation ostensibly through redemption or reincarnation. That is the fabled belief anyway, since we have no proof whatsoever that such places actually exist. There is much confusion about heaven and hell. Nobody agrees because nobody knows. Archaic perceptions of heaven and hell with a residing God and Devil are primitive and unlikely. However, the concept of an alternative unseen coexisting realm is quite plausible. In fact, it is highly likely even from a scientific point of view.

Where does the soul fit into the physics of things? Where in all our universal fabric can our soul possibly exist? In what dimension, direction or vibration will I find my soul nature? Does it have an identifiable realm of its own? Why does the soul have some mystical and magical powers akin to angelic beings seemingly able to shift in and out of our observable reality without space and time constraints? Does our soul nature operate on higher vibrations and/or in alternative dimensions to our physio-biological bodies and minds? What are we really saying when we talk about an interaction between the cosmos and conscious living beings like ourselves, and what relation are we trying to define? Why do we postulate such striking division between us versus the universal makeup? Where and what is such a "spiritual realm" and how much do we really know about it?

Well, it depends on who is asking. A possible alternative coexisting reality has been equated with many concepts such as akasha, fourth/fifth dimension, higher dimensions, quintessence, superstrings, branes, multi-verse, holographic universe, parallel universes, virtual reality, matrix and hyperspace, to mention but a few. Our ability to travel to the "sky" and peer into its "dark areas" with our massive telescopes has discarded the notion that there is a physical heaven out there. However, that does not mean there is nothing out there to discover. Our dimensional micro- and macro-cosmic universe comprises energy and matter, forces and fields, and at least four "dimensions" of space-time. However, a multi-dimensional consciousness comprises even more dimensions yet to be discovered and verified by our sciences. Some of these are no longer mere pipedreams.

Our physical senses are limited to a small range of frequencies of the total electromagnetic spectrum that current technology is able to determine. Higher frequencies outside the range of our sensory perception are ultra-violet light, followed by x-rays, gamma rays and cosmic rays. Lower frequencies are also not observable and consist of infrared light waves followed by heat waves, radar and radio waves. We simply discard anything outside the frequency range our senses are able to handle. The brain itself may play a part in selecting what it wants to interpret and in what manner.

Similar to our limitations of observing the complete electromagnetic range of frequencies is our ability to observe and understand dimensions. Most humans have a simplistic comprehension of the first three dimensions of space, not to mention the yet illusive integrated dimension of time. The moment we analyse the union between our environment and ourselves it becomes blurred and unsure. Anything more than a simplistic consideration of our universal fabric is simply too complex to fully comprehend and utilise by even the most intelligent and competent of conscious human beings. The mere mental gymnastics required to comprehend an extra fourth *spatial* coordinate or dimension, boggles the mind. The expertise of our senses combined with our mental capacity and consciousness are not yet able to comprehend higher dimensions to the full. However, this may change in the future as we increase our knowledge and further our technology to interact with the yet unfamiliar universal fabric and go through more evolutionary changes.

"God" will always reside in the next unknown dimension. No matter how many dimensions we conquer, we will never find "God" in the one in which we find ourselves. "God" will always stay illusive and unreachable. Either "God" chooses not to be part of our quest or perhaps there is no "God" to be found anywhere. Spiritual multi-dimensionality or "heaven" is simply another habitable destination we are trying to master and "God" and/or "heaven" will always stay illusive in the next dimension higher than the one we find ourselves in. Even if "God" can be found in all dimensions, "He/She" is not obvious. Throughout the history of evolution, living organisms became more complex and smarter. As our complexity increases, so does our ability to understand and interact with higher space and time. Humans are the only living organisms to have mastered the conscious use of part of the cosmic fabric. Even so, in higher space than the observable universe we know, we are adolescents. We hypothecate its existence, but we have no idea how to interact and utilise it as yet.

Our experience of physical reality is in three dimensions of length, breadth and height. They are always at right angles to each other. Any reality consisting of space and matter with physical objectivity must extend into all three these spatial dimensions. In our observable and even scientifically verifiable universe, we can only reduce a physical body into the first three dimensions or spatial elements.

There is much speculation about higher space with more complex spatial characteristics. Both scientists and mystics are searching to understand and utilize such space. There is a common perception amongst physicists and mystics alike that there must be a fourth *spatial* component or dimension beyond what we have already discovered. A four-dimensional body in a four-dimensional realm would have four fundamental dimensions, all at right angles to one another. The fourth dimension must therefore also be at right angles to the first three physical dimensions. The analogy can be explained by a point, line, square, cube and hypercube, where every added dimension is at right angles to another. In the "physical" universe that we comprehend there is no such thing as a fourth dimension. It must lie beyond our current physical reality. In this hypothesis, a durational component of time would therefore constitute the fifth dimension.

In primitive animal consciousness, the first dimension is the most obvious. Observation and movement takes place on a one-dimensional level. Whatever is in front of the organism is its most obvious target and sometimes the only one. Through evolutionary ontogenesis, some organisms have conquered second-dimensional consciousnesses. Animals and perhaps even our hunter-gatherer ancestors perceived Earth as two-dimensional or flat. When dealing with short distances our perception of the Earth's surface is flat. To this day, creationists and the "Flat Earth Society" still claim the Earth is flat. Two-dimensional observation and movement takes place in any direction of 360 degrees. The next or third-dimensional observation takes us one step further where we duplicate the 360 degrees horizontal observation to add a vertical 360 degree component. This is pretty much the awareness most humans have today. From about 30 000 years ago, humans started to store food in their back yards rather than hunting and gathering food on the plains. Eventually we created three-dimensional homes with water wells in our backyards rather than living in caves near water. We even looked at the sky hoping to fly like the birds, which eventually we managed to do. Nevertheless, all of this was still limited to the surface and biosphere of the Earth and to the confines of our physical universe.

Physicists and cosmologists no longer consider our universe to exist in only three dimensions of space, as was the case in classical geometry. Einstein incorporated time as the fourth dimension or integrated component of space and this is mostly our perception to this day. However, that was not always the case. Excluding Einstein's perception of time as the fourth dimension, we can also include one or more spatial component beyond the known dimensions of the physical realm. Ever since Einstein, many have discarded the notion of higher space with added spatial dimensions. Even so, many searchers of higher realities have persisted to ponder whether the mysterious worlds sometimes perceived through mystic practices are not perhaps the next dimension. Is there a fourth "spatial" dimensional component undetected by our five senses and modern technology and is it potentially the spiritual dimension? Most mystics and Theosophists still entertain this possibility.

The late 19th and early 20th century perception of the fourth dimension was often equated with our spiritual dimension. Theosophists and most seekers of higher forms of spirituality still see the fourth or higher "spatial" component as the unseen dimension of the spiritual realm. It is important to notice that the occultist view of the spiritual dimension is not simply limited to alternative vibrations only as is sometimes believed. In order to understand or observe a higher component of reality beyond the purely physical, one has to add a fourth spatial component. Mystics have for some time conjectured that extrasensory perception and spiritual vision stem from the ability to access this realm.

In a book called *Theosophy and the Fourth Dimension* (1928), Alexander Horne writes about the esoteric understanding of the fourth dimension and a possible extended spiritual realm. He points out how considerable attention has been given to the esoteric study of the fourth dimension. The purpose of his explanations of the fourth dimension as a spatial component was to present the subject to students of theosophy and the occult. According to Horne, a better understanding of the relation between the universe and our consciousness can be achieved through such a study. Many students yearn for an expansion of consciousness and for greater vision, yet do not know just what these things mean, and in what way they manifest themselves in the mind. Horne maintains that a study of the fourth dimension will bring an expansion of consciousness and a joyful revelation. He sees the apprehension of the fourth dimension as an expansion of consciousness that can unlock some spiritual faculty still dormant in most humans.

Horne reckons we can arouse our four-dimensional faculty into activity through exercises similar to the way one would educate a dormant muscle to function. The struggle to understand the fourth dimension is accentuated when some claim that it is impossible for humans, living in a three-dimensional world and supposedly being a three-dimensional entity, to visualize the fourth dimension. In Horne's experience, many have succeeded in visualizing the fourth dimension. He admits, however, that the understanding of what the fourth dimension represents and implies, still requires a persistent effort of the imaginative faculty.

Horne admits that we cannot physically point out the location of the fourth dimension, we can by analogy only give an idea of the direction in which it lies. For Horne the theosophical and occult perception of the direction of the third dimension is characterized by the word "upward", and the fourth dimension is characterized by the word "inward" or "within". The third dimension is presented as above and away from anything in the two-dimensional universe, and the fourth dimension within and away from anything in the physical universe. Horne says the use of the word "within" and even the way in which we refer to soul "within" stems from this inner fourth-dimensional experience. It all creates a sense of being "inside". Even if we cut up the physical body into a thousand fragments, we would never discover the soul. The fourth dimension is obtainable by analogy from the familiar three dimensions of our physical world, and Horne thinks that the mathematical characteristics of that world could be known, and that the spatial relations between this world and the next can be determined.

To become familiar with more dimensions, Horne alleges a person would, after a while, lose all reality in his mind and that reality would become only an abstraction. We still have not reached the plane of absolute reality, there is still more beyond this. Once we have reached the fourth-dimensional reality, it becomes difficult to live in the limited realm of three spatial dimensions only. Horne points out that we have not yet reached it and it is impossible for the three-dimensional being to understand the four-dimensional reality. A lower being can only see an infinitesimal portion of the superficial universe. Our emotional, intellectual and spiritual life, Horne says, is outside of a three-dimensional being's universe and beyond the limits of such consciousness. The four-dimensional information is still very limited to a three-dimensional being. The three-dimensional being does not know the hand that controls it nor of the force and intelligence that might control it. We are only conscious of the shadows cast by the real things. Since we do not have physical evidence of anything else, we think that what we see is all there is. We do not see the universe as it is, but only as it appears to us.

Horne is convinced we are not conscious of the universe in itself, but rather of the impressions that the universe makes on our perceptive faculties. Our limited perception of reality struggles to make such visualization. The four-dimensional theory shows the universe is not necessarily all there is. From analogy, we see there can be another, higher realm that is not separate from the one we know, but includes it. We see only the surface of things. Our observable universe forms but an infinitesimal part of a greater universe. Horne points out that the forces we see at work in the physical world have their sources in a higher-dimensional reality.

Whether the fourth dimension is spatial, or rather durational as Einstein suggested, does not seem to be a problem for Horne. Even though the two views may seem mutually contradictory, both can be regarded as a fourth component of the space-time continuum. However, if the fourth dimension is an extension in "time" it cannot also be an extension in "space". Horne points out that the apparent contradiction is cleared up when it is realized that there is no dimension in space which can monopolize the designation "fourth". Theoretically, there could be an infinite number of "fourth dimensions". By applying the fourth dimension as spatial, we do so with a wider universe than this physical reality in mind – something Horne calls an "occult world" or "superphysical universe". Horne maintains the extension of duration or time as the fourth dimension can only apply to the physical world itself. Since we do not refer to time and a fourth spatial component at the same time, both can be called the "fourth" dimension. Even though Horne thinks the fourth dimension as a spatial element is equally as acceptable as one that is durational, he admits that it creates plenty of confusion.

Horne maintains the use of the fourth dimension has a useful application with respect to the phenomenon of clairvoyance. As much as any one- or two-dimensional entities such as a line, a flat surface and a square have a higher counterpart in the dimension above itself, similarly we can by analogy reason that our perception of reality is an infinitesimal portion of something bigger. All indications are there is something much more extensive and complete than our physical realm locked up in the spiritual or four-dimensional (spatial) realm. The ability to tap into it allows one to obtain more information. Horne considers the ability to see the fourth dimension as factual and already applied by clairvoyants who ostensibly experience supernormal powers of perception. For Horne the analogy allows us to theoretically deduct the possibility that the observable universe may have a super-physical, higher-dimensional counterpart and that we may be more complete beings than is manifested in this physical reality in which we find ourselves. The fourth dimension seen this way can explain a higher reality as well as the many cases of clairvoyance so frequently attested too. The theory allows us to say it is possible for a person with a hypersensitive nervous condition to contact a point in the four-dimensional counterpart. This will allow one to look into the three-dimensional body from a point in four-space and both entities can be equally visible. However, Horne admits that even though the corroboration from the realm observed is considered fact, it remains a theory.

Lastly, one could perhaps consider Horne's association of the hypothetical four-dimensional world with the "astral world" of occult literature. He thinks the descriptions and characteristics mentioned with regard to the astral world by clairvoyant vision tallies practically in every respect with the description of the spatial characteristics of the four-dimensional universe. The only places where they may differ, is when we have literary difficulties and speech limitations to describe what we experience and occult books describe spatial characteristics in non-mathematical terms. For Horne there is never any real discrepancy and the occult and mathematical descriptions are eventually identical. Horne therefore identifies the astral with the four-dimensional world, and thinks the word "within" exactly corresponds to that world.

It is not difficult to see why occultists see the spatial rather than the durational element of the fourth dimension as the next step up from our observable space. Clarience and our ability to experience higher dimensions are far easier to explain with added alternative spatial components. Until we come up with better explanations of how and why humans experience the esoteric realm as we do, modern science can conjure up as many explanations as it wants to about the non-existence thereof, it will be to no avail. We have to find credible answers to our mystical experiences and psychic phenomena. If not, we may all be delusional and imagining things, which obviously is also possible. However, we cannot discard the possibility that humans may be far more sensitive instruments that outperform the best modern technology has to offer.

Chapter 17

Space-time

The use of modern physics to explain spirituality and a spiritual dimension is widely criticised by the scientific community. All attempts to do this have failed and I have no illusions about my ability to make a reasonable contribution in this regard. That is not the point. What is important though, is that we do not exclude sincere attempts to utilise whatever our physicists discover to assist us in the quest for spirituality. I find nothing wrong with combining spirituality (not religion) and physics in the search for survival. Although combining the two may seem awkward at first, in the context of the exploration of life, it is appropriate. Survival takes precedence over prejudice. Physics will always present new insights, which affect our views on spirituality.

How can physics help us in our quest for spirituality and the soul? How far are we on our quest to alternative realities beyond the physical realm? Can the soul-mind consciousness exist in a higher realm that equates with spiritual? What kind of reality is required for the soul to achieve spirituality? What must physicists discover to advance spirituality?

If anything, science rather than religion will enable the soul to achieve its goal – immortality in another realm. Even if we practice religion to give us hope, we most likely still need science to do the job. If we look at the soul from a non-religious viewpoint, it becomes imperative that science complements the soul's quest for survival. However, not all science or physics does this. Space exploration does little if anything for our soul's quest for survival. The soul does not need space ships where it is going. The soul also does not need a nuclear bomb or a computer for a brain. The question then becomes what scientific mechanisms are required for the survival of the soul nature?

The soul has to survive without a corporeal body. Distinguished astrophysicists Lawrence Krauss and Glenn Starkman raise the possibility that consciousness may exist without the physical body: "*Ultimately, we will need to abandon our bodies entirely. Though futuristic, the idea of shedding our bodies presents no fundamental difficulties. It presumes only that consciousness is not tied to a particular set of organic molecules but rather can be embodied in a multitude of different forms, from cyborgs to sentient interstellar clouds.*" (The Fate of Life in the Universe, *Scientific American Special Edition*, 2002.)

Shedding our bodies is a different concept altogether to discovering new planetary and stellar destinations. At a news conference in June 2006, Hong Kong, the British astrophysicist Stephen Hawking suggested that humans could have a permanent base on the moon in 20 years and a colony on Mars in the next 40 years. According to Hawking the survival of the human race depends on its ability to find new homes elsewhere in the universe. Hawking and other physicists suggest the human race must migrate to other planets and solar systems with space ships. However, this is not what Homo angelicansis requires.

The history of human endeavour to understand space does not date back far. Ancient geometers from Egypt, Sumer and Babylonia needed to calculate right angles for the corners of buildings. In doing this, they realised that all measurements are only possible in three directions. This may have led to the notion that space is three-dimensional. However, modern physics of space and time has its roots in classical and empirical geometry refined and systematized by the Greeks who discovered they could apply such measurements to the rest of the universe. For these early scientists the flow of time was separate and independent of the material objects the world contained. Space and time were each considered absolutes. Until 100 years ago, the flow of time was still thought to take place at an even rate, independent of the material world – a concept modern-day physics refutes.

Zeno, the Greek mathematician and philosopher of the 5th century BC, invented a philosophical argument that illustrates a number of paradoxes relating to space and time. One of his famous paradoxes says that it is impossible for a runner to reach an end goal such as a winning post. From a mathematical perspective, the distance would have to be covered at half the distance at a time. The problem arises when we divide the distance in half, ad infinitum. You would cover half the distance an infinite number of times, which would take forever. We would simply get closer and closer to the finish line without ever reaching it. However, this is not what happens in real life, for a runner is able to reach the finish line even if he does so half the distance at a time.

Similarly, we are faced with a minimum timeframe when movement from one end of a minimum to the other cannot take any time. The time it takes to cover a minimum distance will take a negligible unit of time. Seen in this light Zeno tells us that space and time cannot regress by infinitely small degrees because, eventually, we will reach an indivisible minimum. The concept of space, time and motion is therefore by no means as straight forward as one would think. Modern physics determines that the line cannot merely be the sum of points and time is not merely fractions of a specific period.

Space and time would have to have minimum fractions for us to cross the finish line, or else we would have to find new ways to explain what happens at this level of nature. The true nature of both space and time are still unclear. Amongst other considerations, we do not know whether the nature of space and time are continuous or if you can endlessly divide any distance or duration in half and in half again into ever-smaller units.

According to Max Planck (1858-1947), a German physicist and father of quantum theory, the smallest measurement of length that has any meaning is roughly 10^{-33} centimetres. This minimum measurement, called the 'Planck length', is far too small to detect with any modern technology. Ever since the beginning of the 20th century, the Planck length formed the basis of modern quantum mechanics. Planck also invented the "Planck time", which is 10^{-43} of a second – the smallest unit of meaningful time. Planck time is calculated by the time it takes light to travel the distance of a Planck length. When dealing with currently accepted physics, we run into problems if we explore space and time smaller than the Planck scale.

Planck length and time helps us to make sense of dimensional objects such as the one-dimensional line and the two-dimensional plane if we see it as composed of zero-dimensional points. According to 20th century physics, it is meaningless to try to partition space and time more finely than the Planck length and time criteria. Our universe ceases to have meaning at such minute points of measurement. Nothing can be fitted into such meaningless fractions or have any measurable period beyond such level. Presently the concept of ever-smaller space and time ends at the Planck scale. We still have no idea what happens at this scale and beyond. For space and time to exist at such minute measurements means that new rules of physics must apply – such as extra dimensions. Extra dimensions become the basis of new theories such as string theory whereby more dimensions than the four dimensions of space-time are investigated. (See Chapter 18)

Some classical physicists dating as far back as 500 BC speculated that space was flat. With flat space, if you travel far enough in a straight line, you would fall of the edge of the Earth at some point or you would end up with the stars or dark regions space. But it was also about this time that some Greek philosophers argued that the Earth had to be curved rather than flat. Even though they did not know how to explain why the surface of Earth seemed curved, several observations made them question the nature of space. The way an approaching ship appeared on the horizon made them realise the Earth could not be flat. In the case of a flat Earth, the ship would appear small and get bigger as it gets closer. However, a ship's sails first appear on the horizon, followed by rest of the ship – as if it rises from the sea.

The 3rd century BC mathematician Euclid was largely responsible for the principles of geometry and mathematics. Euclidean geometry is seen as the true nature of space until modern physics overturned this view more than 2 000 years later. Euclid's text *The Elements* represented the basics in geometry until the end of the 19th century. To this day modified versions of his books still form the basis of instruction in plane geometry in junior-grade schools. Euclid wrote five fundamental postulates of geometry. In Euclidian physics, our physical universe has three dimensions of space with a freestanding aspect of time. This kind of space is entirely consistent with everyday experience and with all forms of ordinary measurement of size and distance.

Mathematicians found themselves having trouble with Euclid's fifth postulate or parallel postulate. The fifth postulate states that it is only possible to draw one line parallel to another through a point next to the initial line. This baffled physicists for more than 2 000 years and it is only in the 19th century that answers started to emerge. Since Euclid dealt with flat surfaces, the mathematics of the fifth postulate requires that space must be straight or flat – any line in or through the universe would remain infinitely straight.

Modern-day mathematicians introduced the idea that space could be curved rather than the customary view that it was flat. In the 19th century, mathematicians Karl Friedrich Gauss, Nikolai Lobachevsky, and Janos Bolyai independently expelled Euclid's fifth postulate of the unique parallel. All three mathematicians introduced abstract mathematical ideas known as non-Euclidean geometry to demonstrate that it is possible to have a consistent system of geometry in which Euclid's fifth postulate would be replaced by a new one. Parallel lines can in fact cross one another. This calls for a different type of space – one that is curved. A familiar example is the lines of latitude on the surface of the Earth. They start parallel at the equator, yet cross at the poles. The German mathematician Bernhard Riemann developed the mathematical framework to describe how space is indeed not flat. The heart of Riemann's geometry shows that the spatial dimensions of space can indeed be curved. The weight of a body causes space to warp around it. This would mean that the entire universe consists of curved or warped spaces.

The next advancement toward a better understanding of the fabric of space occurred in the late 19th century with the unification of electric and magnetic forces. British physicist James Clerk Maxwell came up with a theory based on the cumulative efforts of Andre-Marie Ampere, Charles-Augustin de Coulomb, and Michael Faraday. They discovered that electricity and magnetism displayed a close relationship with one another – together they exist as electromagnetic waves that can travel throughout space at the speed of light. Maxwell's observation that electromagnetic waves travelled at the speed of light signalled that light is an electromagnetic wave. The obvious conclusion was also that if light was a wave, it had to move in a medium. It became a controversial point of debate as to whether Aether or ether (mentioned in earlier chapters) is the medium it travels through.

The concept of ether has not been proven in physics. All experimental attempts to detect ether have thus far failed. In 1887, Albert Michelson and Edward Morley were unable detect any form of ether in one of the most significant experiments performed in the last century. The experiments involved measurements of a beam of light. The Michelson-Morley experiments showed it was impossible to detect ether with current technology. Modern physics showed ether was either not detectable or simply not there. The possibility must still exist that something like ether remains hidden in higher dimensions – something not easily acknowledged in those times but now more readily assumed to be true.

The German-born physicist Albert Einstein solved some of the most perplexing scientific problems of our time. He showed that geometry is not intrinsic in nature. Einstein developed two of the greatest scientific insights of all time with his special theory of relativity in 1905 and general theory of relativity in 1915. With Einstein's relativity theories in place, we abandon the flat space of Euclidean geometry and the linear flow of time.

Maxwell's equations were the first to demonstrate how all observers will experience the same light speed. This helped Einstein to create his special theory of relativity. In special relativity, space is no longer conceived as merely three-dimensional – it is intrinsically and inseparably interlinked with time to form a four-dimensional "space-time" continuum. The fourth dimension, however, is not a spatial component but a durational one. Special relativity also shows that there is no preferred state of rest in space in the entire universe. There is no absolute reference for space, time or motion – they are all relative. Observers with differing motion will experience simultaneous events differently. At different velocities, they will also order events differently in time. The relative experience of space-time is not an illusion, but a real phenomenon. In normal daily life, we experience time and space as absolutes due to our relatively slow motion. With increased motion, especially closer to the speed of light, speed becomes an absolute quantity and space-time becomes relative.

In a four-dimensional space-time universe such as our own, time stands still for anything travelling at the speed of light. This means that for a photon or light particle travelling at light speed, time stands still. Some of the insights of special relativity are difficult to process. If time stands still at the speed of light, it would mean it took no time for light to travel from the big bang to the present. The universe has been unfolding and evolving for 14 billion years for observers in the rest of the universe and on Earth, but it took light particles from the beginning no time at all to travel the distance to enlighten us about it. The photon does not experience time at the speed it travels relative to the Earth. Some photons still carry the exact information from 14 billion years ago. By studying light that has been travelling for billions of years, we are able to unravel the marvels of the past. What Einstein managed to do is to establish that the physical universe is not quite as obvious as one would think. Special relativity shows that space is different for all observers in different circumstances.

In general relativity, Einstein goes further to demonstrate that even gravity, the weakest of the four known forces, will affect the way space-time behaves. With general relativity, the structure of special theory is adapted to include gravity. As much as time is intrinsic and inseparable to space, space-time is intrinsically linked to the distribution of matter. The curvature caused by gravity extends to four-dimensional space-time. By adopting Riemann's geometry Einstein was able to predict how gravity indeed curves space and time.

General relativity advocates that the effect of gravitation observed near any mass deforms space-time around it. The degree of curvature depends on the mass of the object. The relationship between space, time and now with the inclusion of matter is relative – it again depends on the state of motion of the observer. General relativity demonstrates how the entire universe is dynamic with varying measures of space-time. With general relativity, it would be impossible to distinguish between being pulled by the gravity of our planet and being accelerated by a lift. Euclidean geometry is no longer applicable. A line is no longer straight but rather the shortest distance between two points – and it does not have to be straight.

Even though relativity does not favour any number of space-time dimensions, ultimately Einstein restricted it to physical bodies in four-dimensional space-time. If we have to rely or depend on four dimensions to secure a spiritual escape, we are doomed. We have to discover science beyond Einstein to have success in spiritual migration. This does not mean we can discard Einstein's insights altogether. Our understanding of relativity is part of our journey towards spirituality. We have to use it as a step towards spirituality rather than a handicap in that journey. Nevertheless, it cannot be all there is.

Even though we know Einsteinian space-time is relative and affected by motion and mass, it remains a four-dimensional continuum. Unless we somehow discover that space-time has more hidden secrets capable of hiding spirituality and a soul, we must still merely presuppose there is more beyond. What will actually satisfy our quest for a spiritual realm? In all likelihood, the search for a soul and a spiritual realm is mere fantasy. However, even if we sometimes venture on futile journeys, we must investigate all possibilities to make sure we do not miss something. For now though, relativity and curved space does not include a spiritual realm. There must be more!

Chapter 18

Extra Dimensions

It is unlikely that the soul is restricted to four-dimensional space-time. If four-dimensional space-time and a physical nature are all life or "God" has to offer, the outcome will be extinction. It is all a matter of time. By the time our sun runs out of hydrogen we must have found ways to migrate elsewhere. But what does elsewhere mean? Presently we have two scenarios – firstly physically colonising other parts of our solar system and perhaps even our galaxy, and secondly spiritual (metaphysical) migration. Both require science we still have to discover. Nothing we have mastered thus far allows for any form of escape.

The question is – will science and spirituality take on a new meaning if we discovered extra dimensions? Is a higher-dimensional realm the "heaven" of both these disciplines? Now that humans have acquired cognition and consciousness, does it become crucial that we discover higher dimensions to ensure its survival? In spiritual terms, there cannot only be four dimensions of space-time. There has to be more. In fact, the first four dimensions of space-time are possibly not even required. It would be awesome if we discovered there are higher dimensions and our souls were already masters thereof. My perception is that our chances of success will increase considerably if we discover and conquer higher dimensions. But is that realistic? Will we be able to discard seemingly futile projects such as our current space travel programmes?

The scientific idea that extra dimensions exist dates back to the 1920s. Its origins are twofold. The first is the discovery that the universe is expanding. In 1922, the American astronomer Edwin Hubble demonstrated that some of the lit star-like spots in the night sky were actually galaxies and our galaxy is merely one of billions scattered throughout the universe. Then, in 1929, Hubble made another fundamental discovery that the galaxies were actually flying apart from each other – the universe was expanding. A second contribution towards an understanding of extra dimensions can be attributed to the Polish mathematician Theodor Kaluza and the Swedish mathematician Oskar Klein. Both of them independently proposed the possibility of a new unseen fourth spatial dimension. With the help of Einstein's theory of relativity, Kaluza and Klein set out to unify the forces of gravity and electromagnetism, which seemed more likely to be explained in a five-dimensional space-time realm. Although the details failed, the introduction of an extra spatial dimension is very relevant. Extra dimensions have since changed the way physicists think about the universe.

Initially the idea of an expanding universe raised the question of what would happen when such expansion eventually runs out of steam. With an energy density more than the critical density, the universe will in time stop expanding and collapse under the influence of gravity. To see if gravity would ultimately halt the expansion of the universe and pull all matter back into a big crunch, cosmologists made numerous measurements to add up the mass in the universe. To their surprise, they discovered the amount of visible and detectable matter in the universe does not add up.

The distribution of visible matter in the universe turns out to be far less than the critical value to explain our expanding universe. It seems there must be some form of undetected "dark matter" in the universe. If this is true, the visible matter of the entire observable universe makes up a mere 5% of the critical value. Ninety-five percent of matter is simply missing or undetectable with our current technology.

Matter of this nature seems to consist of non-luminous material that cannot be directly detected with telescopes or any form of electromagnetic radiation. The presence of dark matter is deduced indirectly from the motions of stellar, galactic and supercluster observations. In August 2006 scientists from the United States claim they may have found the first direct evidence that dark matter exists. By studying the gravitational impact of two colliding galaxy clusters with orbiting telescopes, Douglas Clowe from the University of Arizona and a team of researchers say they have detected dark matter by noting a gravitational pull on the visible matter of these clusters.

For most of the 20th century further in-depth investigations of extra dimensions ended on the shelf. In 1998, however, an even more spectacular discovery was made. Not only is the universe expanding, it is also accelerating. Studies on supernova (exploding stars), reveal that the expansion rate of the universe is not even. Astronomers have noticed that galaxies further away from Earth are receding faster than the neighbouring ones, which means the expansion must be accelerating. Observing the redshift (spectral differences toward longer wavelengths), we are able to determine that supernovae further away from Earth are moving faster than those closer to us. The wavelength of the light from distant galaxies is stretched more due to the expansion of the universe.

To explain the acceleration cosmologists deduced that a form of unknown and as yet undetected energy is causing it. Overwhelming evidence has now established that not even dark matter is enough to explain accelerating expansion. In addition, rather than a universe simply dominated by gravitation, the universe must be composed of an even more alien form of energy. This is where quintessence, discussed in earlier chapters, fits in. Unlike the attractive nature of gravity, this "dark energy" repels matter and causes the expansion to accelerate. Astrophysical measurements now indicate the universe is composed of one-third "dark matter" and two-thirds "dark energy". No one knows for sure what all of this means or what these dark elements are made of. It has become a bit embarrassing for scientists not being able to explain more than 90% of the makeup of the universe.

The idea of such dark energy has long been postulated in history. Dark energy seems to be a constant form of energy representative of the "cosmological term" or Lambda already proposed by Einstein in 1917. Einstein, as everyone else at the time, was convinced that the universe was static. He was unable to construct a static universe with only matter and curvature. The continual gravitational attraction of matter could cause the universe to eventually collapse upon itself. In order to counterbalance the attractive gravitational force, he introduced an ad hoc cosmological constant. Unfortunately for Einstein it was only 12 years later that Hubble discovered the universe was far from static. Hubble's discovery that the universe was actually expanding, forced Einstein to abandon his cosmological term. Einstein's apparent mistake, which he considered to be the "biggest blunder" of his career, had to wait the rest of the century to be revived and given new recognition.

Even though Einstein considered it his biggest blunder and there have been numerous cases for its dismissal, the cosmological constant was resurrected with the predication of dark energy. An expanding universe poses a serious problem in cosmology. We are unable to explain such expansion without adding mysterious unknown elements to make our calculations work. Einstein's "blunder", or his original idea of a cosmological constant, may yet provide the key to solve the dark-energy problem.

The possibility that dark energy originates from the physics of extra dimensions is real. We cannot exclude the possibility that dark matter permeates our entire universe like the akasha – described in previous chapters. The quest to discover the existence and makeup of dark matter and dark energy makes for one of the most exiting challenges of modern science. Some astronomers and physicists have argued that the dark matter may consist of exotic elementary particles completely different from the matter we know. Propositions of minute particles or larger structures for dark matter already exist. These include "neutrinos", "brown dwarfs", "black holes", "Weakly Interacting Massive Particles" (WIMPs) and possibly "white dwarfs". Not for the faint-hearted!

A unified theory able to explain nature and the way the universe is put together has never been successful. One of the problems holding physicists back is their failing to devise a theory capable of combining nature at large with the realm of the sub-atomic level. Relativity is merely one of the two cornerstones of modern physics – the other is quantum physics. General relativity provides a theoretical framework for understanding the universe on a large scale, whereas quantum mechanics provides an understanding of the universe on the scale of sub-atomic particles. As yet the two have never been conjoined successfully to form one ultimate encompassing theory – they remain incompatible.

Quantum physics came to the fore in the 1900's when the German physicist Max Planck postulated the revolutionary idea that matter can only emit or absorb energy in small discrete units called quanta Another profound discovery was made by the German physicist Werner Heisenberg who in 1927 found that it is impossible to specify the position and momentum of any subatomic particle simultaneously This phenomenon is called the uncertainty principle. If you measure the position of a particle, you automatically cause a disturbance in the velocity of that particle. Knowing or measuring the position and velocity at the same time is impossible.

One of the most explored theories in quantum physics is the Standard Model. The Standard Model is a physical theory that explains our current understanding of known or elementary particles and the known fundamental forces of nature on a sub-atomic level. The Standard Model is incomplete as a unifying theory since it only incorporates the electromagnetic, weak and strong forces of particles, but amongst other things it does not incorporate gravity. An additional limitation is its failure to explain the cosmological constant problem. Dark matter is seemingly not made up of the particles the Standard Model explicates. The Standard Model explains only the particles of the observable universe.

The biggest problem we face is the unification of general relativity, quantum mechanics and all these unknown dark elements of nature. To unify our understanding of the universe on both scales, we require something like a Standard Model which includes gravity. It must explain all matter and all four of the known forces in nature on cosmic and quantum mechanical scales. A suitable theory must even go further to incorporate and explain the yet undetected dark matter and dark energy as well. Not a mean task at this point. Physics of this nature is still extremely challenging.

Some new experiments to confirm extra dimensions are in the pipeline. The next 5-10 years may confirm the existence of hidden dimensions. The main contender for discovering such is the 27 kilometre circumference Large Hadron Collider (LHC) at CERN near Geneva which will be switched on in 2007. The European laboratory for particle physics will smash energetic particles together and probe regions never observed before. If all goes according to plan, the physicists may even create exotic types of matter and most importantly find evidence supporting extra dimensions. Even if the evidence is indirect, the expectation that we live in a multi-dimensional realm might be revealed. It seems needless to comment on the tremendous impact such a discovery will have.

One of the most promising constitutions for an ultimate unification theory is "String Theory", also known as M-theory and Superstrings. String theory attempts to marry quantum mechanics with gravity and form a theoretical framework capable of unifying quantum mechanics and general relativity. In the theory, particles and forces are described as tiny oscillating strings. Based on the Kaluza-Klein idea that extra dimensions exist, "strings" extend the physical realm of four-dimensional space-time to permeate a realm of 10, 11 or perhaps as many as 26 dimensions. The string must vibrate in at least 10 space-time dimensions: the four already established by Einstein plus six curled-up in the Planck length. To be mathematically consistent, strings are calculated to be curled-up in a multidimensional realm as small as the unimaginable Planck length. If so, it would almost certainly be too small to detect directly, making some scientists question whether it can be a valid theory. String theory is still in the making and physicists are not entirely sure what the outcome will be. However, the beauty of the theory lies in its anticipated ability to unify quantum theory and gravity. It also challenges the conventional idea that one cannot divide space and time smaller than the Planck length of 10^{-33} cm and at the very least, it opens our eyes to the possibility that space-time exists on more than four dimensions.

Recently theoretical physicists have questioned the assumption that extra dimensions can only be as small as the Planck length. According to influential physicist Lisa Randall, professor of physics at Harvard and professor Raman Sundrum from Johns Hopkins University, an invisible extra dimension might also be big or stretch out to infinity – making it hidden or hard to see. Their work of 1999, known as RS-1 and RS-2, suggests the universe evolved differently in the beginning. Their solution explaining the weakness of gravity is that there must be an extra or fifth dimension. Since there are no known names for such extra dimensions, Randall calls them passages. Randall and Sundrum suggest the existence of an extra dimension could resolve some of the most basic mysteries of our universe. Their papers explain why gravity could be a billion times weaker compared with electromagnetism and the weak and strong nuclear forces. It is possible that gravity is separated from us by five-dimensional space-time. In such a scenario the remaining three stronger forces and the particles stick to our realm and gravity is free to travel across space-time. The possibility that gravity is from another dimension could explain why it is so weak.

21st century physics deal extensively with extra dimensions. It has spiritual written all over it – finally we are getting somewhere. It is mind-boggling to think that physics has only seriously explored extra dimensions in the last 25 years. It is almost as if physics must play catch-up with mystical and esoteric wisdom. What if physicists really do discover extra dimensions? Moreover, what if the hidden dimensional reality is the same one mystics encounter, and similarly the same realm the soul resides in? What would the purpose be to discover higher space if our physical nature and technology are unable to utilize it? There must be a reason and purpose for discovering higher dimensions. Alternative and more successful forms of space travel for one, spirituality for another.

With the advent of space travel and our ability to reach the moon, physicists have become increasingly interested in finding ways to extend such journeys to far-off solar systems. In order to make such journeys possible we need to find a "short-cut". We simply cannot transpose our physical nature to ultra-far-off destinations in four-dimensional space-time. Twentieth century physics simply does not allow for feasible interstellar or intergalactic travel. Even so, the possibility that modern physics will discover and master alternative higher dimensions and "short-cuts" in space is extremely plausible. The definitive question though, is whether the spiritual realm equated with soul life is similar to or the same as the extra dimensions still eluding science.

I find it appropriate that we are discovering extra dimensions at this time of life and evolution. It comes at a crucial time when we are able to re-engineer humans through technology and create cyborgs or worse. If we do not discover higher realities or realms, we are in trouble. The moment we realize or discover extra dimensions, it opens doors with far-reaching possibilities. It gives us more options. Perhaps we can send new beings on each of the two destinations and see which journey ends successfully. Perhaps this is the required split so long anticipated and prophesied about, only it is not a religious destination but a scientific one.

Three things of importance are happening simultaneously at the moment. We are in the process of discovering and defining consciousness, we are set to discover extra dimensions, and we are investigating consciousness outside the physical brain. Is it mere coincidence or are they linked? Is it also coincidence that it comes at a time when we realise how near impossible interstellar travel in a four-dimensional space-time continuum is? Perhaps we are coming to grips with the impossibility of such a task. Perhaps we are about to have a change of direction in the sense that we might look for alternative ways to travel beyond as I suggest in this book. It seems we are looking for alternatives to space travel.

It is highly unlikely that heaven or hell awaits us. However, there may be alternative dimensions beyond our observable space-time universe likened to former perceptions of heaven and hell. If anything, our spiritual nature will find alternative realms for our next evolutionary destination. There is no reason why our consciousness cannot exist in extra dimensions. If higher dimensions exist, we are almost sure to find consciousness already operating there. Perhaps the Homo sapiens' brain is not evolved enough to actualise this.

The difference between Homo sapiens and Homo angelicansis is that Homo sapiens seeks the mastery of space to get to alternative destinations, while Homo angelicansis needs to get into space-time and higher-dimensional space itself. Homo sapiens wants to get "there" and colonise new destinations with the physical body intact. Homo angelicansis merely wants to get "there" irrespective of the physical body. The hypothesis of extra dimensions to which a soul must migrate is therefore not that far-fetched after all.

The existence of extra dimensions makes it possible that we are surrounded by parallel universes without even knowing it. This hypothesis suggests our spiritual nature can reside in a parallel universe undetected. Having a shadow soul in a neighbouring dimensional realm is possible and perhaps even probable. There is no reason why we cannot be part of an extra dimension if it exists.

One thing is sure, if we want to become spiritual, we have to conquer higher dimensions irrespective of which method is used.

PART 5

New Frontiers

Chapter 19

The Spirit of my Soul

Science does not acknowledge informative life-giving principles such as prana, chi or spirit. In addition, there are no etheric and astral realms as supportive realities to our physical nature. Even though life and the sequel of evolutionary events tell a story, modern science does not think such grandeur was premeditated or created, and its storyline was not designed, directed or choreographed. Science cannot give credit to a "force" it cannot measure or a "God" it cannot find. This inability to grasp what makes matter appear alive, impels scientists to discard the notion that something supernatural is running our evolutionary universe.

To modern science, organic matter with the anatomy of a living organism has no maker and it does not require any help or assistance from anyone or anything to form and maintain life. Living organisms originated spontaneously and the successful reproduction of offspring to form part of the evolutionary process happened spontaneously. We do not know whether intelligence or supernatural spiritual forces or perhaps even a miracle from a "God" was a prerequisite for any such physio-chemical building-blocks to come to life and become a living organism.

How then can we have two virtually identical cells next to each other, one alive and the other mere lifeless matter? What differentiates dead matter from living matter? Embryonic stem cells have the ability to divide endlessly and have the power to turn into any type of cell. How stem cells "know" in advance what cell to become remains a mystery to science. Evolutionary science determines that all known life is the result of a single or typical phenomenon when matter formed a structure with life potentiality. The way in which atoms (matter) organized to form a physio-chemical structure with life potentiality in the form of a cell, happened purely by chance. Similar structures managed to stay alive long enough to pass its blueprint on to a succeeding generation and eventually this process was repeated, changed and improved endlessly and without fail to this day. What makes atoms group together to form a particular cellular structure and what makes various cells "stick together" and appear alive in any particular coherent form, is not entirely known. However, modern science determines that there is no supernatural or hidden life-giving principle that makes life and its diversity possible.

Steven Rose, a professor of biology, in his book *The 21st Century Brain* explains a possible scenario of how this unassisted formation of life on Earth may have started some four billion years ago, and how it evolved to where we are now. He points out that the A, C, G and T, which is the alphabet of DNA, comprise life's alpha and omega for many present-day molecular biologists seeking an origin of life. However, for Rose the inception of life or the beginning thereof starts first with primitive cells with something we might call behaviour. Taken from a variety of sources, Rose has assembled a convincing scenario which explains everything from the origins of cells and the formation of genes right to the present day brain and its cognition. He points out the relevancy of what he calls behaviour, that he thinks should be considered before the constitution of brains and even before nervous systems existed. Behaviour is the goal-directed actions by cells upon the external world.

Rose explains that we need chemistry to begin with. Life, however defined, engages the complex interactions and "interconversions" of chemicals built from carbon, hydrogen, oxygen and nitrogen. These molecules float in a watery sea with the ions of sodium, potassium, calcium, chlorine, sulphur, phosphorus and a range of heavy metals.

The first problem is to move from the inorganic chemistry of our cooling earth into the organic, carbon chemistry of sugars, amino acids and nucleotides. In turn, they become the building-blocks of nucleic acids, and from there the giant molecules such as proteins, fats, DNA and RNA become possible, bringing phosphorus, sulphur and the rest in along the way. Even though there are several ways in which this could have happened, the simplest way was possibly through synthesis in places like the oceans, volcanic eruptions or perhaps in clay-like parts where land and water mingle. He points out that some even suggest life may have resulted from outer space. It would probably also have required something like fierce electrical storms to get the initial process going. Rose explains that abiotic synthesis theories like these are not entirely exclusive and if only one or all of these scenarios have been in play the result is that Earth ended up with organic chemicals possibly covering the oceans and margins.

To proceed from a lifeless layer or pockets of chemicals to a point where life is contained is still a big step. Rose considers the answer may lie in the lipids or oils of the ocean. Oil on the surface of water will either spread as a thin film or form a droplet. These droplets have a physio-chemical property that can concentrate many of the organic chemicals within themselves. A lipid membrane effectively creates an inside and an outside, which is the basis or the structure for the environment of any potential organism. The droplet becomes what Rose calls a proto-cell. The proto-cell has an internal chemical constitution that is different from what is on its outside. This separation is the primary characteristic of any living organism. The difference between what is on the inside and that which is on its outside is what has the potential for self and non-self. Rose points out more complex chemical interconversions are possible once the separation of organics has taken place. Computer models can show how mutually interacting chemicals are able to eventually stabilize and form a dynamic metabolic web. Larger organic molecules are also able to form within such a web. Eventually strings of amino acids or nucleic acids (RNA) are able to catalyze their own synthesis, enabling them to accelerate or even direct further syntheses.

Rose explains that the ionic composition inside the proto-cell is involved in catalytic processes whereby the internal concentration of the ions becomes different to that outside the cell. A simple lipid membrane will not achieve this specificity for it cannot readily select what may enter and what it must exclude. In living cells, however, the semi-permeable membrane is selective and chooses which ions or molecules may enter or leave. Rose considers that in some way a variety of proteins in the lipid environment must have been incorporated into the lipid proto-cell membrane at a relatively early stage in life's history. This configuration for life potentiality is fundamental in the start of a living organism. It is actually possible to measure the difference in voltage on the inside and outside of living cells and membranes. Two electrodes, one inside the membrane of a modern living cell and a second on the external surface or in the surrounding fluid, connected to a voltmeter, will record a significant voltage difference between the inside and the outside of the cell. The inside of the cell is negative and the outside positive. Rose points out that the membrane potential is as defining as any other aspect of cell structure and biochemistry. All proto-cells need energy for internal stability and survival. They also need energy to grow and divide. There is a sufficient amount of energy in the oceanic environment in the form of synthesized molecules such as molecules of sugars. Eventually these proto-cells acquired the capacity to capture the sun's energy and tap the atmospheric supply of carbon dioxide. It would also have used solar energy to synthesize sugars through photosynthesis to generate and release oxygen. Billions of years later, the Earth ended up with an oxygen-rich and carbon-dioxide-poor atmosphere, as we know it today.

Rose continues to explain that faithful reproduction becomes the next significant step in evolutionary development. By sucking more and more material in, proto-cells become larger to a point where they lose their stability and split into two. However, primitive reproduction from splitting up does not yet represent faithful reproduction or replication. For similar offspring cells, the same pattern of proteins must be present in each. This, Rose explains, is where the mechanism for faithful reproduction starts to play its role. For accurate synthesis of complex molecules where each daughter cell is identical, we require RNA and DNA. RNA exists as a single-stranded molecule and in DNA two strands are locked together to form a double helix. DNA strands have the ability to separate and form a template for another strand. However, there is still uncertainty whether early replication is based on RNA or DNA. Rose notes that there is a paradox in this account of the origin of life. Enzymes and energy are required to synthesize a nucleic acid molecule from scratch. In addition, controlled energy production requires enzymes and cellular mechanisms. Yet, cells as we know them today depend on DNA for their accurate synthesis. In Rose's opinion, for proto-cells to turn into fully-fledged cellular organisms, there must have been some sort of bootstrapping. Some self-help situation, whereby the fundamental molecules involved in cellular energy trading are closely related chemically to the molecular building-blocks from which the nucleic acids are assembled. Nevertheless, as Rose points out, this view is unresolved and still only based on well-informed speculation.

Cells with higher efficiency at capturing and using energy and of replicating more faithfully have a far better chance of survival than less efficient cells, which would tend to die out. Rose explains how two evolutionary processes take place simultaneously. The most common evolutionary process established by Darwin is based on competition, where two rivals struggle for survival. The second process, which is less acknowledged, is based on co-operation between organisms. Rather than competing against one another, cells can also team up to work together.

This brings us to one of the most important stages where all the defining features of life are in place. Cells are capable of metabolism and faithful replication. There is symbiogenesis and competition. There is the presence of a semi-permeable boundary separating self from non-self and it has the potential for metabolism to self-repair and maintain this self. Lastly, the potential for reproduction of more or less faithful copies of this self is in place. For Rose all of these features require a condition he calls adaptability or behaviour, which is "*the capacity to respond to and act upon the environment in such a way as to enhance survival and replication*" (*The 21st Century Brain* – p 22.) It is this behaving principle that Rose calls living. At its simplest, behaviour does not require brains or a nervous system. What it does require is something like a programme (Rose uses the word cautiously). The programme must be flexible enough to adapt to short- and long-term eventualities of the external environment. One way of considering the capacity to vary a programme is some form of action that needs to take place to achieve a desired goal. For long-term survival, it must be sustained at least until replication is achieved.

Somewhere in the evolutionary history, cells started teaming together into larger aggregates. The ability for cells to group together and find mutual benefit is profound. "Multicellularity", as Rose calls it, had a profound change in life-style. The merit of social living is that it is no longer necessary for each individual cell to respond adaptively to rapid changes in its environment. The survival of an individual cell becomes dependent on the survival of the organism as a whole. In a multi-cellular setup, you have to find ways to co-operate through communication. Cells can also specialize – a condition called homeostasis. However, Rose prefers to use the term "homeodynamics", which emphasizes how stability is achieved through dynamics rather than through stasis. According to Rose, behaviour becomes a property of the organism as a whole in a multicellular system, whilst the needs of individual cells are subordinated.

In order for cells in a multi-cellular biotic community to communicate, they would require modes of communication. For a direct line of communication, a nervous system is required. According to Rose, neurons must have evolved from secretory cells. With such a system, messages could be conveyed electrically or chemically – both phenomena are known to occur. With increasingly complex forms of behaviour, various specialized cell types evolved to where we are now.

Ultimately, these multicellular organisms evolved to have brains with the capacity of cognition. According to Rose, our brains did not only evolve to make humans the smartest animals on Earth, they continue to be concerned with survival, better strategies for survival and differential reproduction for our offspring to have a better chance of survival. Both co-operation and competition are better achieved with improved brains and even though brains develop cognitive skills, it has to be cognition for a purpose. For Rose the purpose of cognition is also surpassed by what he calls emotion – animals learn and remember for a purpose that includes effect. Rose feels strongly that information is empty without a system to interpret it, to give it meaning. Animals with brains have hormones that interact directly with their brains. Rose explains that learning always involves some form of emotion, and emotion engages more than just the brain. Ultimately, we ended up with Homo sapiens and in Rose's opinion, our uniqueness over and above all other animal species lies in our versatility. Even though humans are outclassed in virtually every department by other animals, we happen to be the pentathlon experts. Humans have also developed many other facets such as language, consciousness, foresight and we have other unequalled structures of society, culture and technology. We are also the only animals capable of telling or writing about our achievements.

The most common perception for the elicited arrangement of life is the belief in miraculous acts from a "Creator God". "God" represents an omnipresent being ceaselessly upholding the entire creation with some kind of spirit. In occultism, prana constitutes the spiritual life principle and etheric bestows form on life. Together with matter, they determine biological life and form. In turn, astral forces determine the etheric principle and so forth. What determines all is not certain, so we have to revert to "God" – or whatever name we like to bestow upon such an entity. Even if there are life-giving, etheric and/or astral principles in a supernatural spiritual realm to inform the physio-chemical and physiological nature, we still have to explain how it manages to administer life and create the variety of function and form we have on Earth. How do the etheric and astral principles obtain their abilities and mind?

Let us assume for the moment that there are higher principles and that the spiritual realm does exist. What possible scenario would we be faced with? Let us also assume that there is some sort of mandatory life-giving force for organisms. If the first ever structure of matter with life potentiality came to life some time in the past and died soon after, it would mean the end of the entire "life cycle" of life itself. Life existed as long as the first "alive" cell lived. The premature death of the first cell would have put an end to any potential for the continuation of life itself and for evolution into higher organisms that may have resulted from its existence. If a life-giving force supported the "life" in this biotic organism, its potentiality, too, came to an abrupt end. If the first life form managed to replicate itself, it passed the life of its body on to the next generation. In such event, even if the parent organism dies, life itself can carry on its existence in the offspring for as long as it manages to keep its body intact and nourished. Living organisms have no power over this life-giving force. For some reason, any living organism must sustain most of its material form to stay alive. Even though the living organism is able to change through growth and aging, we cannot alter the physical composition of the organism significantly without the matter losing its life principle. Each living organism's lifespan differs. Life stays with any organism for a relatively short period, and then it permanently ceases to exist. However, life as a whole proceeds.

As much as we require life to exist, life needs us to exist too. As far as we know, the spontaneous creation of life from nothing with the continued ability to replicate was only required to succeed once. Against all odds, one tenacious archetypal ancestor cell managed to come alive and somehow managed to replicate itself. All that was required thereafter was to imitate the achievement of the first successful archetype. Nevertheless, all resulting life had to take place without any interruption or break. At no point was there, or can there ever be, a break where no carrier of life exists, otherwise life in its entirety will cease to exist, and there is no guarantee that this phenomenon can or will ever repeat itself. Not only must each complex organism have life to exist, each of its individual sub-components right down to the cells, atoms and the sub-particles of the atoms must either continuously behave collectively as a living organism or act as if it has this informant. The repeated and sustained cycle of life produced the increasingly more complex sequence of selective evolutionary events and continues its efforts to this day. As long as at least one cell in this universe manages to maintain "life", the cell and the life donor has a chance of continued existence. Imagine the incredible magnitude of atoms required to fall into a perfect sequence to form the anatomy of only one cell. Then imagine how all these atoms, in unison, simply behave as if alive. Yet our sciences determine that matter required no additional informative force to make such continued animation possible.

Any potential life-giving force is representative of the informative principle giving life to a chemical structure to become a biological structure. For those who still hold on to the possible existence of a "ghost in the machine", the ability of "life" to appear directed and choreographed is representative of its spirit. The animation that living matter demonstrates makes it appear to have an informative force and/or spirit. Our inability to comprehend the phenomenon of "life" in matter makes us question the existence of a life-giving principle. The existence of informative forces and/or an instructive consciousness for living organisms has never been verified from a medical or scientific perspective. Due to this lack of evidence, modern science is of the opinion that all living matter came about by chance and, through evolutionary adaptation, is able to endure on its own.

The inability to make sense of "life in matter" is the reason why religious doctrines attribute the mythical status of "Holy Spirit" to it. Even so, neither any such life-giving force, nor any life-giving spirit, is the soul. From an esoteric viewpoint, each individual living organism as well as each dissident species carries the same "gift of life" so to speak. Without it, you seemingly perish. Assuming that gift of life does exist, a life-giving force that all living organisms appear to have, clearly cannot be the soul. The soul is not this giver of life. The soul must be something else. The soul, like all of us seeks life, it does not give life. For billions of years no organism had a soul, but all had life – spirit. The soul, if it exists, followed.

Does the soul also need "life" to exist? Once our physiological body dies, can the soul exist without a higher life-giving force? I have showed how the soul seeks life and that it seeks ways to maintain our consciousness after physiological death. However, does the soul too require "life" to exist? Is the soul just another aspect that forms part of the cycle of life? Is the soul merely a body of another kind? Does the soul have its own life-giving force to give it life for continued existence? In all likelihood it does. The soul appears to be the ultimate product of the long and arduous process of evolutionary life and the struggle for survival. The soul may be the supreme outcome of selection in evolution. The soul did not make the entire journey from the beginning of life; if anything, it resulted towards the very end. The role of the soul becomes far clearer when we look at it in this manner. The soul cannot be a life-giver. It does not give us life and it will not give us life when we die. If anything, it too will seek life to sustain its animation.

Somehow, we think that the soul is capable of independent immortal existence after the body and its mind dies. Is such a perception realistic? Is there a quantum leap separating the triune nature of brain (body), mind and soul? Is each of the above separate and freestanding? If separate, by implication, the soul could have originated somewhere else and is capable of an independent existence or journey without the body. Its immortality was achieved before it joined this physio-biological journey, before joining the Earthly experience, and the soul can continue its journey without the mortal body and its temporary mind.

Alien souls incarnating into human embryos in the wombs of pregnant mothers pose immense problems (Chapter 3). The belief that an alien soul is a separate entity which accesses and leaves the body before and after the Earth being comes alive and when it dies, only to reincarnate again at a later stage, has never been validated. By saying there are two separate entities, we are surmising they influence each other but are not responsible in equal measure for the other's fate. Whom do we hold responsible? There is no judicial system on the planet that will allow such notions. How can we say the body and its brain-mind cannot be held responsible for the independent decision-making of the soul? The law that holds an individual responsible for his or her actions can no longer be applicable. If I have a body and a brain-mind as well as a separable soul-mind, who is responsible, who gets the blame and who is punished? If the soul cannot be linked to the body and the brain-mind, why do we reward or punish the soul for the deeds of the body and its brain-mind? They are either one and treated as such, or else we have to see them as separate and hold each as an independent unit. These are but some of the implications of arguing that souls are separate and separable entities capable of autonomous and immortal existence. The dilemma of this argument is typically demonstrated in religions such as Catholicism and Christianity. Their doctrine teaches that the soul is punished for the sins of the body and yet can be rewarded through salvation.

The contrary view must be that that the three – body, mind and soul – are all part of one organism and we have a gradual transition from one to the other. In other words, they are connected parts of a whole (being). The soul stems from the living organism. It is my view that for the soul to exist it must be connected and yet not entirely separable and immortal as most religious doctrines would suggest. If they are part of one entity, the soul still dies when the body and brain dies since the brain-mind and soul are still linked. We are by no means sure yet that the soul manages to survive physiological death. Perhaps the soul is merely an illusive evolutionary component of our being, seeking to separate itself from the body in order to become immortal somewhere in the future. By the time the physiological body dies, the soul also dies for it is still connected to and part of a physical nature. When the life that beholds it leaves the physiological body, in all likelihood the soul dies too.

If the body, mind and soul are connected, then the soul is locked in a struggle for immortality. If triunism exists, the body, mind and soul are one, and all are interconnected in a way we cannot yet determine. Our failure to determine how they are linked is merely the result of physiological and cognitive capabilities and our deficient technology. The soul as a vehicle is therefore part of the organism's nature and possibly still mortally linked to the physiological structure. The soul would have to evolve to a point of independence and once it does so, it would have to rely on its own autonomy to exist. That autonomous existence may also require the gift of life, whatever that means. Even though the soul is not of this dimension, it still is part of what makes up the whole organism. Part of me also exists in a higher dimension. Part of my nature co-exists in hyperspace. It is not someone else or some alien soul invader; it is merely part of me. I do not think it is separable and immortal yet. However, I do think it is on its way to achieving this goal.

Chapter 20

Transmigration

Apart from the many interpretations of spirituality already discussed, there is another form of spirituality to be mentioned. Some people spiritualise without formalising their experience into any pious, esoteric or philosophical context. It is a kind of spirituality based on a belief system and expressed through a sensitive quality for life represented by a state of being spiritual, a place of peace and harmony. This form of spirituality does not necessarily imply religion; it is merely the desire some humans have to love and to love life and be in touch with the goodness in people and nature. It is an undefined experience that expresses a desire to be "spiritual". To me this kind of spirituality is a desire to be good and love every creation in order to make the world a better place and to try to eliminate what seems to bring about evil, pain and suffering. Nevertheless, this form of conviction is not the kind of spirituality I am explaining here.

The animal kingdom, which we consider lower on the evolutionary ladder than humans, has no experience of good and evil. When a lion kills a deer, it is part of nature. The rest of the fleeing animals do not think in terms of how evil the predator lion is for killing one of its kind, making this deed punishable by death. It is simply part of nature. In the case of complex conscious creatures such as ourselves, a division of good or evil has been defined and some choose to express their spiritual experience with kindness and generosity to lesser animals and to fellow humans rather than to do them harm. This kind of spirituality is not really a scientific one in the same manner as I am describing and exploring in this book. The spirituality I am dealing with is not merely an emotional expression of generosity towards all creation, but an evolutionary change that takes place where a specific part of our nature becomes spiritually different.

In order to define spiritual anthropology I have removed the concepts of belief and faith from my studies. We have no need for these forms of conviction. Either we know something or we do not. The things we do not know yet, we must discover and validate. All truth, in principle, is relative but the validity of anything unknown does not require faith. It requires insight, knowledge and substantiation. Faith or belief is indicative of the measure of ignorance we have about a subject. We do not rely on faith when we present theories to the world; we give credible explanations of observed physical and cosmological phenomena. We have to realise possible answers to problems and present it to the world in the form of theories and hypotheses, not as doctrine. To validate our theories we have to substantiate it with experiments. If these are accurate, it constitutes proof that at least, under certain conditions, our theories are correct.

Our scientific search for spirituality and the soul is similar. Either the soul exists or it does not, but there is no need for faith or belief to make spirituality or the soul plausible. For the soul to exist, we should be able to verify its existence scientifically. We can know that the soul exists and whilst we still cannot verify its existence as factual, there is no need for faith or belief. We simply do not know everything there is to know about it yet. In the same way science at times points in certain directions without making complete sense, we have to find explanations and credible answers. Desire to know the soul does not require faith in dogma.

Neurobiology tells us it is impossible to have mind and consciousness without using the brain, and yet people whose brains have been clinically dead for 40 minutes were able to display cognisance during this time. Their observations take place from a vantage point outside the parameters and confines of their physical body. According to modern-day physics, it is impossible to predict future events and yet some people are able to see and prophesy the future with uncanny accuracy. Such visionary insights are achieved without the use of any scientific methods of observation. According to modern sciences, we should merely be able to make predictions based upon experience and predictability. Nevertheless, some people do display "supernatural" powers that appear prophetic.

Are these the doings of a spiritual persona in a spiritual realm yet to be discovered? These seemingly impossible observations contradict all current understanding of the physical laws governing our physical universe. Nevertheless, we should be able to find credible answers for such "supernatural" abilities without reverting to magic or mysticism. Until such time, these mysteries merely fall outside the spectrum of our current understanding and knowledge. That is all!

Homo angelicansis implies evolution has shifted from the physical body to the spiritual soul. For some time evolution was concentrated above the neck as the human brain took a quantum leap and in the process achieved consciousness. Evolution has now made yet another quantum leap with the advent of the soul. Why would human evolution do that?

Intergalactic Migration

Earth as we know it is running out of space and resources. All life on Earth is dependent on sufficient real estate and natural resources for its future survival. Eventually Earth's land, fossil fuel, food production and reserves will be maximised, making any further exploitation impossible. This problem is exacerbated by the ability to prolong life indefinitely. We are trying to utilise the entire earthly biosphere and its resources to maintain our physical existence and biological life. In some parts of the world, we have already reached critical parameters causing the extinction of thousands of species of fauna and flora. Earth's climate is changing as a result of the carbon dioxide we release into the atmosphere through the burning of fossil fuels. Scientists have now determined that many plants and animals are threatened by the effects of global warming. According to some studies, the climate change is becoming the most serious threat to the planet's biodiversity. In some respects, we are even selectively causing the extinction of our own kind. That which natural disaster and catastrophe does not kill, we do.

Humans are looking for new frontiers. As conscious beings, we control our environment to the point where it is impossible for other life forms to survive without our consent. Eventually even humans are simply going to run out of everything we depend on for physical survival. Scientists are hard at work to find ways to sustain life beyond Earth. We have already found the moon uninhabitable. Mars is one of the only planets in our solar system that we think may have had life or may be able to sustain life. However, Mars does not have a large enough mass and gravity to produce a full biosphere to make it inhabitable for humans. It also appears that Mars lost its core-driven magnetic field, a shield that would have to protect it against the fierce pounding of particles from the Sun. This currently prevents the planet from having a carbon dioxide atmosphere or watery oceans.

Beyond our own solar system, the distances are so immense we will only be able to reach the nearest stars in hundreds of thousands of years at the highest possible speed current space travel has to offer. We know of distant planets in far-off solar systems, but they are not yet a proposition for we cannot tell if they can sustain life. We cannot see planets in nearby solar systems with our strongest telescopes. They merely reflect the light of the nearby stars but do not shine enough, making it impossible to observe them directly at such long distances. We observe these planets by studying the gravitational effects, which cause the sun in their own solar system to "wobble" as it goes around in its orbit.

We have no feasible technologies that we can use to explore neighbouring solar systems, let alone alternative galaxies in the rest of our universe. It is still so improbable it is futile to discuss the mathematical models for such space travel here. Conditions suitable for anthropomorphic life require a planet with enough pressure to maintain water in the liquid state, gravity large enough to keep atmospheric pressure, and the correct distance from a star for suitable temperatures. Even if we found another planet with such conditions, it will still take hundreds of thousands of years to get there and probably millions if not billions of years to make a lifeless planet look and function like Earth. It took evolution on Earth more than three billion years to make it suitable for human life. Our choices are to find a planet that is already suitable, change other planets to suit our needs or change our own nature to be able to survive space travel and the conditions of alternative destinations.

The scientific conjecture for future survival is based upon the preservation, transportation and resettlement of our biological nature to alternative destinations through inter-stellar space travel. The keen interest of science fiction writers and Hollywood filmmakers over the last century has been to show how we can utilize futuristic science and technology in unique ways to allow inter-stellar space travel. This is made possible by manipulating our own physical nature to harness the nature of space and time itself. Unfortunately, it is not as simple as making movies using our overly keen imaginations and a fair share of artistic liberty. Even if we based our movies upon scientific theories, they are still only pipedreams of what we would like to achieve once we become superhuman.

Thomas Georges, author of *Digital Soul*, explains how he anticipates our future survival. To live longer for these intergalactic journeys he wants to replace our existing bodies with new ones and maintain the conscious part of the original ones. Georges notes that even though we have the technology to extend our own life span by controlling disease and by replacing ageing organs with ones borrowed from other humans or even mechanical ones, only a few are able to live longer than 100 years. He wonders whether we will eventually be able to extend life indefinitely by replacing more worn-out body parts with synthetic substitutes, to a point where we end up as Homo cybernaticus.

Georges goes further to explore the possibility that we might as well replace an entire worn-out body with stronger, better-designed, and longer-lasting counterparts that merely support the brain. The idea is to rid ourselves of our biological bodies altogether, but keep the brain intact. Georges believes there are two seemingly distinct paths to immortality; one is to supplement and eventually replace our organic parts and perhaps eventually even our brains, which can become mechanical, the other is to design and control the configuration of our organic bodies and minds from scratch. He thinks it is likely that the biological and mechanical paths will eventually merge and we should one day be able to engineer any kind of organic modifications we wish, which could include entirely new human beings.

Georges illustrates his scepticism when he expresses concern about the moral dilemma of playing God. If we do this, where will we allow ourselves to go before we stop? Being able to design human beings what will prevent us from making greedier, violent, barbaric, self-absorbed humans instead of nicer, testosterone-free, superhuman beings? We may even decide to make entirely different kinds of intelligent life altogether. There seems to be no limit to what we can do. This, Georges says, is where the difference between extinction and immortality becomes blurry.

Georges certainly has a point when it comes to the possibility of creating completely new or different kinds of beings. This is exactly my point with Homo angelicansis, in that we are becoming completely new beings without the biological bodies we are accustomed to.

According to Armand Delsemme, retired professor of astrophysics at the University of Toledo in Ohio, the next step in the future of human design is more optimistic. In *Our Cosmic Origins* he explains how genetic engineering is plausible and perhaps even evident irrespective of current religious, moral and ethical objections. With the awakening of intelligence, the rules of the game change completely. He believes we are beginning to use molecular biology for profit. Initially we will restrict ourselves to the correction of accidental genetic defects, but later we will modify the human genome in directions yet unforeseen with uncertain results. This we will do despite ethical, religious or cultural considerations,

Delsemme notes that our DNA already has a repair capability that gets rid of copying errors in the genome. By improving this there is no reason why humans will not be able to extend the survival of our species by decreasing the number of useless mutations. His question, though, is whether it would be desirable to do so, since the opposite of selecting rapid evolution of our genome by favourable mutations could lead to a kind of "superman". He calls for deeper knowledge of the human genome that can offer new techniques and open up unimaginable prospects. Delsemme is also aware of the risks of moral, social, religious and cultural conflicts. In the end, he thinks our different cultures will adapt quickly to changes in the long term.

For most parts, it seems as if Delsemme is focusing his ideas on changing the physical part, which is mainly genetic, and the mechanical part of our nature as previously described. He agrees that we do not know enough about artificial intelligence to explore all the possibilities at this point in time. We can perhaps replace dissipative structures based on carbon chemistry with new dissipative structures based on something like silicon and make organisms more intelligent than we are. However, the main objective is to change the fate of a civilization confined to a single planet with very few chances of long-term survival. Delsemme hits the nail on the head when indicating that one of the most important reasons for changing our nature is for long-term survival, which must eventually determine our ability to spread throughout the galaxy (or beyond).

Humans are becoming so advanced technologically that we will be able to lengthen the average life span and perhaps even be able to sustain life for hundreds of years. However, this will not get us very far in cosmic terms. To facilitate inter-stellar travel we will have to sustain life for hundreds of thousands of years and more. If we are able to prolong life nearly that long, will not evolution be obsolete?

John Gribbin, author of books such as *In Search of the Big Bang* (1986) and *Science: A History* (2002) points out how important research on time travel has become. Mathematical physicists in the 1980's were more and more convinced that time travel was possible and that it lay within the spectrum of the laws of physics. Einsteinian physics and in particular the general theory of relativity offer ways of explaining how extreme space-time travel can be harnessed and even conquered some time in the future.

During the 1980's, the main contender for space-time travel was via tunnels in space connecting two black holes called wormholes. Another theory was based on massive rotating cylinders – a form of time machine built in space, capable of creating enough drag on space-time through its spin to make space and time travel possible. Both theories are based on the fact that space and time are not definite entities. These are simply theories and currently have no future or even long-term solutions. It may be that in time we will be able to utilize or discover ways to harness these methods. All this remains in the realm of science fiction rather than applied science.

Houston, we have another problem! We cannot stay in space for long enough to go far enough. Our biological bodies are not yet capable of surviving the harsh environment outside our biosphere for long enough to make space travel possible. This is compounded by the lack of a clear-cut and workable space-time travel programme capable of propelling these bodies (our biological bodies and space ships plus the fuel to propel both) away from the Earth's gravitational field and into the unknown abyss of space.

Jeffrey Sutton of NASA's space medical research team explains that our physical bodies are very vulnerable when subjected to even relatively short journeys in microgravity. The human body as it is today shows severe bone loss at a rate of as much as 1 to 1,5% a month. This means space travel leads to an acceleration of age-related changes similar to osteoporosis. Extended exposure to weightless environments such as outer space effectively increases the risks of kidney stones and bone fractures, which are related to bone demineralization. Astronauts also experience loss of muscle mass, strength and endurance. Microgravity affects the neurovestibular system, which is an integrated set of neural sensory and brain circuits that allow humans to maintain balance and directional orientation and stabilize vision. The body also suffers loss of blood volume, immunodeficiency and transient post-flight anaemia or low red blood cell levels, even if adequate nutritional intake is maintained. Space radiation poses one of the greatest risks to humans on prolonged missions. Radiation can induce cataracts and cancer, as well as adversely affect many physiological processes. There are perhaps possibilities that through genetic engineering we may be able to create super-humans that will be able to withstand such harsh space travel without disastrous effects on the body, but then we still have the physics of space travel to deal with.

Gregory Stock in his book *Redesigning Humans* also acknowledges that Homo sapiens is not the final word in primate evolution and believes we have actually come to the realization that we are on the cusp of profound biological change. To contemplate the notion that humans might become more than "human", he says, seems preposterous since we are still biologically identical in virtually every respect to our cave-dwelling ancestors. Never before have humans been able to alter themselves biologically in such meaningful and predictable ways through genetics. Bio-ethicists and scientists have great concerns about the consequences of future genetic technologies, since it is not yet clear what implications new developments may have for reproductive biology.

According to Stock, emerging technologies will soon become the cutting edge of human biological change, and developments of this nature will write a new page in the history of life when these technologies allow humans to seize control of future evolutionary development. Biological enhancement will take us into yet unexplored domains, which will challenge our basic ideas about what it means to be human. He rightfully points out that it is no longer a matter of if, but rather when, where and how humans are going to manipulate embryos.

In his book, Stock examines the emerging reproductive technologies for selecting and altering human embryos. Developments in germline engineering to modify future generations will have huge consequences, he comments. Already it is possible to manipulate specific genes in an embryo to prevent disease. More advanced techniques for germline engineering that will institute more comprehensive changes already call for ethical and social questioning. We have no idea yet where this development will ultimately take us, once it becomes safe and humans begin to apply it to self-design. All we know is that it will transform evolution in that reproduction will become highly selective and much faster and effective at spreading successful genes. Human cloning has started and we no longer need to wonder whether it is going to happen.

Stock mentions how this advancing self-transformation could change and transform future generations to such an extent that we could end up completely different from our present-day selves. He says such an occurrence would more aptly be termed pseudo-extinction and we can expect it to result in the end of our lineage. Homo sapiens would then spawn its own successors by fast-forwarding its own evolution.

In Stock's words, we are deciphering the "code of codes", "reading the book of life", and looking at the "holy grail of human biology". What concerns me is the next step in Stock's argument. He is of the view that some disaster might derail our technological advances or our biology might prove too complex to rework. In my opinion, this is the essence of the dilemma. How capable are predominantly scientifically minded humans to recreate the richness of life that ensures long-term survival and/or overcome future catastrophes?

Have we covered all there is to cover in our assumed ability to play God? Have we managed to consider all it takes to redesign humans through genetic manipulation? Do we have to consider our spiritual nature if we have one? Is our spiritual nature physical even though we call it spiritual? For me the error lies with the view that the entirety of our human genome has been unravelled allowing us to modify our biology. If we look at it on the purely biological basis of only what we can see, we probably have unravelled all there is, but I think there may be more. However, Stock rightfully points out that we are unlikely to achieve the technology to travel around the galaxy without altering our own biology as well. He is also correct in saying the Human Genome Project is only a beginning. That much we do know.

Travel in space does not only have restrictions regarding our physical bodies, but also has limitations on any other physical bodies irrespective of whether they are made of flesh and blood, a silicon replica or whether a machine-computer aided cyborg. All of them are subject to failure in science fiction type space-time travel. Sorry, it still will not work! Why? Because of the mass factor!

Travelling faster and further in space requires energy, and lots of it. To make such journeys remotely possible we have to travel close to the speed of light, and to do that, we have to use infinite amounts of energy. More mass means more energy and more energy means more mass. It all becomes a vicious circle, which ultimately ends in money. In any case, we are currently not even close to the knowledge and technology required to make such trips possible. Even if we do acquire the expertise, we still have the problem of physical bodies that cannot handle extended space flights. Perhaps we also forget that only a select few genetically engineered "cybernoughts" will be able to go on such migratory journeys. It would be impossible to send more than a select few to alternative destinations in space ships. What about the rest of humankind that stays behind?

Homo angelicansis

There is one alternative though: to limit or get rid of the mass factor. No mass may be nature's way of assisting us to make such journeys possible: we simply have to lose weight.

Spiritual Migration

To claim we are souls does not mean that soul comes from "God" or that we come from an angelic realm. However, it may mean that we are on our way to one. If we cannot migrate with our physical bodies intact, why do we not leave them behind? If biological evolution has ended, can we transform on a non-physical level? Perhaps we can evolve to become spiritual beings with the formidable characteristics accredited to angelic beings. We have to become angelic. Our new nature of the soul as a spiritual astronaut is bridging the gap science finds impossible. We already seem capable of consciousness without the physical body. If higher space exists, we can and may already co-exist in it.

Our bodies function as transporting vehicles for minds and consciousness. Since our bodies are physical, it means that we can only move around in a physical dimension, one that we measure in four dimensions of space-time. The movement of our physio-biological nature is still immensely restricted no matter what modern science may master. However, if we extend our journey into a "spiritual realm" of "higher space" the criteria changes dramatically and the need to employ alternative incorporeal bodies becomes evident. The nature of our physical bodies does not allow us to journey into higher space. However, if we can transfer our consciousness to the soul we may have a perfect vehicle for both inner and hyperspace migration.

Theologies have speculated about the possibilities of heaven and hell and our souls residing in higher spiritual realms for millennia. All that time various shamans, mystics and prophets speculated on breaking from the confines of our reality into higher vibrations and dimensions. These "parapsychologists" have meditated on a higher space to which no one can point but which is all around us and in us. Physicists and mathematicians can already calculate it, but no one can point to such higher dimensions yet. Even so, there seems to be no reason why extra-dimensional realms do not exist. Higher dimensions lie at the very heart of the most significant modern day theories that describe the very fabric of our universe.

We are discovering that our universe does not only consist of what we can see or perceive with our five physical senses. Our awareness of "more out there" elicits us to find ways to make journeys possible that can take us there. We are discovering that part of our nature has the right characteristics and abilities to take us to new destinations with completely different characteristics. Call that vehicle the soul if you like. Through meditation, rituals and mind-altering substances humans can already make part of the journey to alternative higher space and multi-dimensional realms. But, finding ways to describe these journeys and destinations is often difficult because our language and understanding is limited. To explain the alternative realms we have to use the language of mysticism or the limiting language of outdated geometry and physics. Both are unsatisfactory in describing the soul's abilities.

Journeys into this realm make us realise how much more complex space and time are than our physical realm. Space and time take on new meaning and sometimes even become meaningless as if not to exist at all. This new destination is sometimes incorrectly labelled as the fourth or fifth dimension. In reality, it has very little if anything to do with the actual geometrical inclusion of a single succeeding dimension. The only reason for explaining it this way, is that we do not yet understand the nature of this spiritual realm. It is simply still a spiritual reality where only a few pioneers have been able to venture. For some time now, many have described the journey as the spiritual body accompanied by a secondary consciousness in the astral dimension.

The journey and destiny have various layers and depths and how far we travel depends on the ability of the investigator (explorer). Part of the journey takes us past scenery that is quite similar to our current physical existence, but the further we go the less familiar the environment becomes. At some point, nothing seems to resemble our universe of space, time and matter. At the farthest reaches, we actually end up at a place where only pure consciousness seems to exist and even universal thoughts are meaningless. Our universe with space, time, matter and life as we know it is not obvious in that realm.

How far can we go and what will we find when we get there? Nobody has complete insight into that question yet. The reason is that you eventually get to a point where you have to leave your biological body behind and it has to die. The point of no return. There are various religious and esoteric interpretations of what we can expect on the other side, but we simply do not know for sure. We have no proof anyone can exist or survive there. This does not prevent us from taking bold steps and adventurously investigating this new destination. Adventure is a definite aspect of our human nature. If there is a hidden destiny, be sure that we will discover and explore it and we will learn how to survive there. It is in our blood, or shall I say it is in our soul.

In the end, what are we trying to achieve? We are attempting to transcend and maintain our consciousness in alternative multi-dimensional destinations without the physio-biological body. The problem is that science postulates that human consciousness is only locked up in the brain. The ideal is to have less restrictive vehicles to travel with. Higher space is more complex and demanding than our familiar "physical" space-time continuum. Our physical bodies with senses and our minds are still limited to these three dimensions of space and one of time. However, our alternative spiritual bodies are entrenched in alternative realms of which we are currently becoming more aware.

What makes our soul-nature so unique is its multi-dimensionality. Unlike our etheric and astral bodies that are largely of an alternative vibratory nature to our physical bodies, the aptitudes of the soul seem far more accomplished and also reside in higher dimensions. In order to understand the functioning of soul's vehicle for consciousness, we have to understand its environment. If our physical brain-minds want to understand the information our soul-minds present to us, we have to understand its other dimensionality and learn to discern its language.

Chapter 21

Revelations

The time has come to rewrite Revelations. The last book of the Bible colligates the end times and Christian eschatology. The scriptures portray the end times and how human existence on Earth will come to end. The author of the apocalyptic doctrine is credited to the Apostle John, canonised as a divine saint by the Catholic Church. Perhaps John's wonderful storytelling may come true, but somehow I think we may see a different end to our future. Two thousand years after the biblical version of Revelations, we are able to look beyond such delusional attempt at prophesy and discover new and far more befitting revelations for human destiny.

Human consciousness cannot be as highly developed without the help of the amazing brain and its network of neurons, which ultimately is a product of evolution. The manner in which our consciousness evolves makes us question whether evolution and change started taking place on a spiritual level as well. Bipedalism and a larger brain became necessary to assist us with new physical tasks in a changing environment. Now that we have an advanced consciousness, somehow we think we can maintain it even without the physical body, hence the need for a soul. Either that or we will change our physical nature through technology and science to assist our bodies in longevity and durability.

In composing this work on spiritual anthropology and soul evolution, I have shown that nobody can be sure that we have separable and independent souls. It is also not sure yet that our spiritual experience is real or merely imagined. Perhaps we are all delusional about a spiritual existence originating in our mythological past. However, when we extract the supernatural and religious mysticism we can consider spirituality scientifically. Can we survive physiological death and maintain some form of consciousness in an immaterial spiritual form? Can the soul transmigrate, reincarnate and sustain life in the manner described by religions and occultism?

The body still dies when we destroy the brain and we are in no manner able to permanently separate consciousness from the host body. Nobody has ever been able to return from such finality to tell the tale. Even though clinically dead patients have had near-death and out of body experiences for short periods, we are by no means sure that it was really a separation of consciousness from the physiological structure. We have only come close to perhaps finding a spiritual realm which exists in the form of higher dimensions, but unfortunately, we have not made the return journey yet. Not even combined efforts between all disciplines investigating the soul have provided unequivocal proof of its existence. We have found no proof that there is any higher destiny other than basic scientific and evolutionary principles already established, and even they are still mere questioning theories.

Where do we go from here? We do what we have been doing for the past 3,8 billion years. It is all about survival of the fittest and sustaining life and consciousness. In order to do so we need to know what we are dealing with now that we are able to participate in playing "God". We could say time will tell, but perhaps that in itself is our biggest dilemma. Is there still enough time to ensure our future or are we going to destroy the very nature we need to support and nurture before time runs out? Is our destiny ending up in the hands of the unenlightened? Are those who assume the responsibility capable or are they about to destroy our only chance of long-term survival? Are we dependent on the survival of the separable and conscious soul? Are we going to harness the skills to make a quantum leap into the next dimension of reality and conquer living in this new environment? Are we going to achieve separation of consciousness from the material host?

Genetic manipulation is a reality and nobody on Earth can stop it, however great the consequences and implications. We are at the crossroads where absolute materialism will triumph over absolute spirituality. Are we going to become Homo cyborgensis or are we going to become Homo angelicansis? We are at a crossroads and every human being has to make that choice. Our time is up.

Which route will ensure long-term survival of our current species? No other time in the history of life on Earth has been more critical in terms of determining the future. It has become a choice. Our conscious experience has surpassed our instinctive ignorance and our technology can already determine varying outcomes. The path towards a soul creature entails becoming a new type of being. That in itself implies a new species, one without a physical body as we know it today. The destination where we say the soul exists is unlike the realm we know. The new destination cannot accommodate our physio-biological nature in its current form. The very least requirement is that we find new "vehicles" to make the journey possible.

If evolution has become too slow to keep up with human intellectual development and if we are going to take the reigns from a "Creator God" or perhaps nature's evolution, can we discard the possible existence of a soul? If we assume the role of "God", are we up to the task? Somehow, I do not think so. Hypothesising about travel to the next dimension is no longer a problem. Nevertheless, the mode of transport is not mechanistic. Over and above sustaining life in the physical form, for some time now, we have been adding to our nature what seems to be a soul. We cannot explain this step in any other way than an evolutionary step to take us to the next destination. Consciousness beyond the boundaries of the physio-biological body offers an opportunity for long-term if not eternal survival.

There does not seem to be any reason to migrate to another dimension yet, since we are presently quite successful survivors. We see everything as challenges, and mastering them becomes a meaningful reward in itself. Earth, for those who can afford to live here, is not such a bad place. As long as all is well, migration to alternative dimensions still seems unnecessary. Nevertheless, what if imminent catastrophe threatens our extinction? We would probably think differently. All our wealth, pleasure and play are worthless in the face of extinction.

This is exactly what is happening to us. We face unavoidable catastrophes. Some catastrophes are looming in the short-term and others are bound to happen in the distant future. This is where evolution comes into effect. We will evolve irrespective of whether we agree to it or whether we are conscious of it. The force or "spirit" of evolution takes place without our consent. It is the most basic and fundamental principle of nature that until now has been beyond our ability to understand, direct or dictate. Only since the end of the 20th century have some humans become able to play a small role and wilfully redirect evolutionary changes. We do not yet know all the implications our input as role-players will have. For this reason, we question the authority and ability of those who assume the role of dictators and creators of evolution. Until now, consciousness was not a prerequisite for evolution. Now that we are aware of evolution and simultaneously capable of genetic manipulation thereof, can we allow a handful of scientists to dictate the future of our own and any other species through prescribed evolution? Are they really capable dictators of the evolutionary process? Survival of the fittest in itself implies there are many options – the fittest amongst many. Nature manages to provide such variety. To provide such variety in genetics is simply not possible. Genetics would therefore rely on a predeterminism of the outcome of the experiment.

Although we associate the soul with life after this life, or life between these kinds of lives, my concern is also *life during life*. In other words, what is the soul up to whilst we are living here on Earth? What is its purpose and function from the day we are born to the day we die? If we speed up evolution, can we take the soul with us? In the Western sense, the concept of death is often seen as the final step in a person's life. Rarely do we think the soul has a function throughout life, that our soul is part of the blueprint of our nature to carry us further.

Some believe the soul goes to its final destination and others believe the soul goes into hibernation and returns through transmigration or reincarnation, perhaps undergoing some training in between. To my mind, the soul is the most recent addition to our being and even more recent is our conscious awareness as a species. Consciousness is so new that we struggle to define it. We have created many fields of study to understand the soul too. Like the body and its accompanied brain-mind, our soul is looking for ways to survive, to communicate, transcend and perhaps have fun. It seeks ways to survive the future and even death. Simply put, it desires pleasurable immortality.

If we think about the soul as part of the evolutionary process, it does not have a final destination or an interim destination but is merely the next evolutionary destination. In our changing universe where life struggles for survival, it makes sense that consciousness would try to evolve in a dimension where it is less threatened by natural disaster. To be spiritual or spiritually gifted is therefore to have the ability to live consciously both in the physical and the spiritual realm simultaneously as if in one reality. Conscious awareness is not limited to the minds' cognisance, but includes an acute awareness of spiritual nature as well.

The problem we face is that we say it is only consciousness or soul consciousness that can make this step. In a way, it means we assume all other species have no such luxury of a conscious soul and cannot have the luxury of transmigrating to the alternative hyperspace. However, shamans and mystics who have journeyed to alternative destinations find that they do encounter the "spirit" of these creatures there. All are tapped into the same forces of nature that makes all life possible. All creatures' existence stems from this alternative higher dimension. We must therefore be careful to assume that we are the only beings able to have one foot in the door of higher dimensions of existence. We may perhaps again only be the most advanced beings tapped into that realm. On the other hand, perhaps we are not!

We do not know what makes cells and RNA or DNA come to life and evolve into complex life forms. Chance happening seems too unbelievable, so we attribute this process to a higher being. However, when we assume we have choice it is hard to understand where destiny fits in. Some would argue that even our choices were destined. To re-evaluate and find new answers to the spiritual dimension of existence we need to look at discoveries made long after the fundamental ancient scriptures were recounted. Perhaps similar principles apply and perhaps "God" does not exist in the next realm. Perhaps there is another spiritual realm beyond even that one and once again, we have to transmigrate to another more elevated realm. Being human in a four-dimensional realm is but a phase and once accomplished we continue to evolve on other planes.

We see various splits in the evolutionary family tree. The most recent was when the genus Homo separated from Australopithecus. Before that, a split separated primates from Australopithecus. The next step may well be when the two groups, Homo sapiens and Homo angelicansis, separate to go their own way. Steiner in *Occult Science* maintained that by the very laws of evolution an older faculty of soul loses its full significance when it develops new faculties. As we become adapted to the new, there is no further purpose in hanging on to the old. For Steiner there are individuals who are able to supplement the newly acquired faculties of intellect and feeling with the fully conscious development of higher powers of cognition, whereby they can penetrate into the world of soul and spirit. Those who contributed to explore the supersensible worlds in this new way are able by their own imagination, inspiration and intuition to gain knowledge of the higher regions of existence.

Homo angelicansis

Homo angelicansis is a new species developing alongside Homo sapiens. By surmising that Homo angelicansis is the more spiritual species of the two carrying the lineage that has evolved more advanced souls, we assume the balance of the Homo sapiens species are not as spiritual. Homo sapiens is therefore impaired and handicapped due to inferior or absent souls. In the same way Neanderthals came to an abrupt end some 30 000 years ago when challenged by the Cro-Magnons, "modern" humans may suffer a similar fate.

It is therefore not merely the spiritually religious that are of this lineage, but the spiritually evolved. Spirituality is attained through evolutionary means and not through doctrine and belief in fictitious religions. To be spiritual, therefore, is not to be religious, but multi-dimensionally evolved. The big question is whether we are able to maintain and preserve such spirituality in a new scientific age with all its emphasis on maintaining only the physical persona.

Homo angelicansis underlines a future generation who will achieve advanced spirituality allowing them to transmute into an alternative dimensional existence. It requires a shift in consciousness by harnessing immortality on spiritual levels enabling us to transmigrate and eventually leave our physiological bodies behind. We can describe this new species as a transformed soul-being using a physio-biological body as its host but striving towards a higher, more agile body with higher capabilities and less physical restrictions. It requires a new host or even possibly a complete autonomous existence without any physical host whatsoever. The product could be an immaterial entity capable of transmigration or migration to higher realms.

In the beginning of this book, I said spirituality is not a formalized religion or a doctrine but a destiny. As we now see, that destiny has nothing to do with faith or the belief in some religious creed. We can explain our destiny scientifically.

Irrespective of how far evolved and developed we are now, we still have the capacity to enlighten ourselves further. Our evolutionary purpose and destiny has become a conscious one and survival requires two choices. Either we find ways to migrate to alternative planets and solar systems in space ships, or we find ways to transmigrate to alternative dimensions of reality we currently call spiritual. On the other hand, perhaps the two disciplines will merge to make migration possible, but I doubt it. Perhaps, as pioneers, we will find ways to harness the astral dimension of quintessence and find our elusive soul genes to successfully manipulate them and hasten spiritual evolution.

As for genetics, we need to find the genetic soul configuration in higher dimensions and explore its possibilities. As for the higher dimensions, we must learn from those who already have a foot in the door and investigate its depths. We should stop thinking of alternative seekers of spirituality outside the confines of religious dogma as heretics or evil-doers. Alternative spiritual exploration is necessary for the study of our future and a possible next destination. Those who make in-roads into this type of exploration are the pioneers.

The spiritually evolved and illuminated ones are those who can already make that journey. Congratulations to them. Now we need to learn how to make the journey a permanent one, one that transcends death. Once we take that step, mortality becomes obsolete – in its physiological form, that is. Survival is not a game – it is for real. It is in our most fundamental core nature and life's most primal "spirit" of behaviour. After that comes pleasure. Once we have found ways to survive, we can enjoy our achievements.

Bibliography

Adams, Fred, *Our Living Multiverse, A Book of Genesis in 0 + 7 Chapters*, Pi Press, 2004

Allman, William F, *Apprentices of Wonder, Inside the Neural Network Revolution*, Bantam, 1989

Armstrong, Karen, *A History of God, From Abraham to the Present: the 4000-year Quest for God*, Vintage, 1993

Asimov, Isaac, *Asimov's New Guide to Science*, Penguin, 1987

Audi, Robert, *The Cambridge Dictionary of Philosophy*, Second Edition, Cambridge, 1999

Baggott, Jim, *A Beginner's Guide to Reality*, Penguin Books, 2005

Banchoff, Thomas F, *Beyond the Third Dimension, Geometry, Computer Graphics, and Higher Dimensions*, Scientific American Library, 1990

Barbieri, Marcello, *The Organic Codes, An Introduction to Semantic Biology*, Cambridge University Press, 2003

Barrett, William, *Death of the Soul, From Descartes to the Computer*, Oxford, 1987

Barrow, John D, *The Constants of Nature, From Alpha to Omega*, Vintage, 2002

Baumann, T Lee, *God at the Speed of Light, The Melding of Science and Spirituality*, Are Press, 2002

Behe, Michael J, *Darwin's Black Box*, Free Press, 2006

Bendit, Lawrence J and Bendit, Phoebe D, *The Etheric Body of Man, The Bridge of Consciousness*, Theosophical Publishing House, 1982

Besant, Annie, Leadbeater, C.W., *Thought-Forms*, The Theosophical Publishing House, 1969

Blackburn, Simon, *Think, A Compelling introduction to Philosophy*, Oxford University Press, 1999

Blackmore, Susan, *The Meme Machine*, Oxford University Press, 1999

Blavatsky, H P, *The Secret Doctrine, The Synthesis of Science, Religion and Philosophy*, The Theosophical Publishing House, 1938

Boslough, John, *Stephen Hawking's Universe*, Fontana, 1984

Brockman, John, *Science at the Edge*, Weidenfeld & Nicolson, 2003

Brockman, John, *The Next Fifty Years, Science in the First Half of the Twenty-First Century*, Weidenfeld & Nicolson, 2002

Brockman, John, *What we Believe but Cannot Prove, Today's Leading Thinkers on Science in the Age of Certainty*, Free Press, 2005
Bruce, Robert, *Astral Dynamics, A New Approach to Out-of-Body Experience*, Hampton Roads, 1999
Buchanon, Mark, *Small World, Uncovering Nature's Hidden Networks*, Phoenix, 2003
Buhlman, William, *Adventures Beyond the Body, How to Experience Out-of-Body Travel*, Harper San Francisco, 1996
Calder, Nigel, *Einstein's Universe, A Guide to the Theory of Relativity*, Penguin, 1990
Calder, Nigel, *Einstein's Universe, The Layperson's Guide*, Penguin, 2005
Callender, Craig and Edney, Ralph, *Introducing Time*, Icon Books, 2002
Cannon, Alexander, *Sleeping Through Space*, The Walcott Publishing Co, 1939
Capra, Fritjof, *The Hidden Connections*, Harper Collins, 2002
Capra, Fritjof, *The Tao of Physics, An Exploration of the Parallels between Modern Physics and Eastern Mysticism*, Flamingo, 1992
Capra, Fritjof, *The Turning Point, Science, Society and the Rising Culture*, Flamingo, 1983
Carter, Rita, *Consciousness*, Weidenfeld & Nicolson, 2002
Casti, John L, *Paradigms Regained, Unravelling the Mysteries of Modern Science*, Abacus, 2000
Cavendish, Richard, *Mythology, An Illustrated Encyclopedia*, Time Warner, 1992
Cayce, Edgar, *Modern Prophet, Four Complete Books*, Gramercy Books, 1990
Chown, Marcus, *The Universe Next Door, Twelve Mind-blowing Ideas from the Cutting Edge of Science*, Review, 2003
Clancy, Paul, Brack, Andre and Horneck, Gerda, *Looking for Life, Searching the Solar System*, Cambridge, 2005
Clegg, Brian, *Light Years, An Exploration of Mankind's Enduring Fascination with Light*, Piatkus, 2001
Collins, *English Dictionary, Third Edition*, Harper Collins, 1991
Collins, Andrew, *Alien Energy, UFO's Ritual Landscapes and the Human Mind*, Eagle Wing Books, 1994
Collins, Andrew, *From the Ashes of Angels, The Forbidden Legacy of a Fallen Race*, Bear & Company, 2001
Cotterell, Arthur, *Encyclopedia of World Mythology*, Parragon, 1999

Coveney, Peter and Highfield, Roger, *The Arrow of Time, The Quest to Solve Science's Greatest Mystery*, Flamingo, 1990

Cremo, Michael A and Thompson, Richard L, *The Hidden History of the Human Race*, Bhaktivedanta, 2002

Darwin, Charles, *The Origin of Species*, Gramercy Books, 1979

Davies, Paul and Gribbin, John, *The Matter Myth, Beyond Chaos and Complexity*, Penguin Books, 1991

Davies, Paul, *About Time, Einstein's Unfinished Revolution*, Penguin Books, 1995

Davies, Paul, *The Edge of Infinity, Beyond the Black Hole*, Penguin, 1994

Davies, Paul, *The Mind of God, Science and the Search for Ultimate Meaning*, Penguin, 1992

Davies, Paul, *The Origin of Life*, Penguin, 2003

Delsemme, Armand, *Our Cosmic Origins, From the Big Bang to the Emergence of Life and Intelligence*, Cambridge, 1998

Devlin, Keith, *The Millennium Problems, The Seven Greatest Unsolved Mathematical Puzzles of Our Time*, Granta Books, 2005

Dunne, J W, *An Experiment with Time*, Faber and Faber, Mcmlviii, 1934

Farthing, Geoffrey A, *Theosophy, What's It All About? Revised Edition*, Theosophical Publishing House, 1983

Ferris, Timothy, *Coming of Age in the Milky Way*, Vintage, 1988

Fisher, Joe, *The Case for Reincarnation*, Grafton Books, 1986

Fisher, Len, *Weighing the Soul, The Evolution of Scientific Beliefs*, Weidenfeld & Nicolson, 2004

Flanagan, Owen, *The Problem of the Soul, Two Visions of Mind and How to Reconcile Them*, Basic Books, 2002

Fontana, David, *Is There An Afterlife?: A Comprehensive Overview of the Evidence*, O Books, 2005

Fraser, Gordon, *The New Physics for the Twenty-First Century*, Cambridge University Press, 2006

Gardner, Laurence, *Bloodline of the Holy Grail, Element Books*, 1996

Geddes & Grosset, *Classical Mythology*, Gresham Publishing, 1995

Gettings, Fred, *Encyclopedia of the Occult, A Guide to Every Aspect of Occult Lore, Belief and Practice*, Rider, 1986

Gleick, James, Chaos, *Making a New Science*, Abacus, 1987

Gleick, James, *Genius, Richard Feynman and Modern Physics*, Abacus, 1992

Gott III, Richard J, *Time Travel in Einstein's Universe, The Physical Possibilities of Travel through Time*, Weidenfeld & Nicolson, 2001

Greene, Brian, *The Elegant Universe, Superstrings, Hidden Dimensions, and the Quest for the Ultimate Theory*, Vintage Books, 1999

Greene, Brian, *The Fabric of the Cosmos, Space, Time, and the Texture of Reality*, Allan Lane, 2004

Gribbin, John and Cherfas, Jeremy, *The First Chimpanzee, In Search of Human Origins*, Penguin, 2001

Gribbin, John, *In Search of Schrödinger's Cat*, Corgi, 1987

Gribbin, John, *In Search of the Big Bang*, Corgi Books, 1992

Gribbin, John, *Schrödinger's Kittens and the Search for Reality*, Weidenfeld & Nicolson, 1995

Gribbin, John, *Science, A History 1543-2001*, Penguin, 2003

Grossinger, Richard, *Embryos, Galaxies and Sentient Beings, How the Universe Makes Life*, North Atlantic Books, 1994

Hamer, Dean H, *The God Gene*, Anchor, 2005

Hancock, Graham, *Flooded Kingdoms of the Ice Age*, Penguin Books, 2003

Hancock, Graham, *Supernatural, Meeting with the Ancient Teachers of Mankind*, Century, 2005

Hawking, Stephen, *A Briefer History of Time*, Bantam Press, 2005

Hawking, Stephen, *The Universe in a Nutshell*, Bantam Press, 2001

Hey, Tony and Walters, Patrick, *The New Quantum Universe*, Cambridge, 2003

Hick, John, *The Fifth Dimension, An Exploration of the Spiritual Realm*, One World, 1999

Horne, Alexander, *Theosophy and the Fourth Dimension*, Theosophical Publishing House, 1928

Hughes-Warrington, Marnie, *Fifty Key Thinkers on History*, Routledge, 2001

Icke, David, *Children of the Matrix*, Bridge of Love, 2001

Icke, David, *The Biggest Secret*, Bridge of Love, 1999

Illingworth, Valerie and Cullerne, J P, *The Penguin Dictionary of Physics*, Penguin, XX

James, Ioan, *Remarkable Physicists, From Galileo to Yukawa*, Cambridge, 2004

Johnson, George, *A Shortcut Through Time, The Path to the Quantum Computer*, Vintage, 2003

Joseph, Frank, *The Destruction of Atlantis, Compelling Evidence of the Sudden Fall of the Legendary Civilization*, Bear & Company, 2004

Kaku, Michio, *Hyperspace, A Scientific Odyssey through Parallel Universes, Time Warps, and the Tenth Dimension*, Oxford, 1994

Kaku, Michio, *Parallel Worlds, The Science of Alternative Universes and our Future in the Cosmos*, Penguin, 2005

King James Version, *The Holy Bible*, American Bible Society, 1981

Klimo, Jon, *Channeling, Investigations on Receiving Information from Paranormal Sources*, North Atlantic Books, 1998

Knight, Christopher and Butler, Alan, *Civilization One, The World is Not as You Thought it Was*, Watkins Publishing, 2004

Koch, Christof, *The Quest for Consciousness, A Neurobiological Approach*, Roberts and Company, 2004

Krauss, Lawrence M, *The Physics of Star Trek*, Harper Collins, 1995

Krauss, Lawrence M, *Hiding in the Mirror, The Mysterious Allure of Extra Dimensions from Plato to String Theory and Beyond*, Viking, 2005

Landsdowne, Zachary F, *The Chakras & Esoteric Healing*, Samuel Weiser, 1986

Lawton, Ian, *Genesis Unveiled, The Lost Wisdom of our Forgotten Ancestors*, Virgin, 2003

Le Poidevin, Robin, *Travels in Four Dimensions, The Enigmas of Space and Time*, Oxford, 2003

Lerner, Eric J, *The Big Bang Never Happened*, Simon & Schuster, 1992

Lewis, Ralph M, *The Conscious Interlude*, Supreme Grand Lodge of Amorc, 1978

Ling, Trevor, *A History of Religion East and West, An Introduction and Interpretation*, MacMillan, 1974

Magueijo, Joao, *Faster than the Speed of Light, The Story of a Scientific Speculation*, William Heinemann, 2003

Marshall, Ian and Zohar, Danah, *Who's Afraid of Schrödinger's Cat?*, Bloomsbury, 1997

Marshall, Reverend Alfred, *The R.S.V Interlinear Greek – English New Testament*, Samuel Bagster and Sons, 1978

Martin, Walter, *The Kingdom of the Cults*, Bethany House, 1985

Mayr, Ernst, *What Evolution Is*, Phoenix, 2001

McCutcheon, Mark, *The Final Theory, Rethinking Our Scientific Legacy*, Universal Publishers, 2004

McDowell, Josh, *Evidence that Demands a Verdict, Historical Evidences for the Christian Faith, Revised Edition*, Christian Art Publishers, 1930

McFadden, Johnjoe, *Quantum Evolution, Life in the Multiverse*, Flamingo, 2000
Microsoft Corporation, *Microsoft Encarta 1993-2005*, 2006
Millar, David, Ian, John and Margaret, *The Cambridge Dictionary of Scientists, Second Edition*, Cambridge, 2002
Narby, Jeremy and Huxley, Francis, *Shamans Through Time*, Thames & Hudson, 2001
Newton, Michael, *Destiny of Souls, New Case Studies of Life Between Lives*, Llewellyn Publications, 2002
Newton, Michael, *Journey of Souls, Case Studies of Life Between Lives*, Llewellyn Publications, 2002
Oerter, Robert, *Mr Tompkins Gets Serious, The Essential George Gamow*, Pi Press, 2006
Oyle, Dr Irving, *Time, Space and the Mind*, Celestial Arts, 1976
Paine, Thomas, *The Age of Reason*, Carol Publishing Group, 1974
Panek, Richard, *The Invisible Century, Einstein, Freud and the Search for Hidden Universes*, Fourth Estate, 2005
Paraphrased, *The Living Bible*, Tyndale House, 1971
Parnia, Dr Sam, *What Happens When We Die, A Ground-Breaking Study into the Nature of Life and Death*, Hay House, 2005
Pearsall, Judy and Trumble, Bill, *Oxford English Reference Dictionary, Second Edition*, Oxford, 2002
Peat, David F, *Superstrings and the Search for the Theory of Everything*, Abacus, 1988
Penrose, Roger, *The Emperor's New Mind, Concerning Computers, Minds and The Laws of Physics*, Vintage, 1990
Powell, Corey S, *God in the Equation, How Einstein Became the Prophet of the New Religious Era*, The Free Press, 2002
Randall, Lisa, *Warped Passages, Unravelling the Mysteries of the Universe's Hidden Dimensions*, Harper Collins, 2005
Rees, Dai and Rose, Steven, *The New Brain Sciences, Perils and Prospects*, Cambridge University Press, 2004
Reichenbach, Hans, *The Philosophy of Space and Time*, Dover, 1957
Revised Standard Version, *The Holy Bible*, Oxford University Press, 1953
Richards, Steve, *The Travellers Guide to the Astral Plane*, The Aquarian Press, 1983
Rohl, David M, *The Lost Testament, From Eden to Exile: The Five-Thousand-Year History of the People of the Bible*, Century, 2002
Rose, Steven, *The 21st Century Brain, Explaining, Mending and Manipulating the Mind*, Jonathan Cape London, 2005

Rucker, Rudy, *The Fourth Dimension*, Penguin, 1986

Rupp, Rebecca, *Four Elements, Water, Air, Fire, Earth*, Profile Books, 2005

Russell, Bertrand, *History of Western Philosophy*, Routledge Classics, 2005

Rutherford, Ward, *Shamanism, The Foundations of Magic*, The Aquarian Press, 1986

Sabom, Michael, *Light and Death, One Doctor's Fascinating Account of Near-Death Experiences*, Zondervan Publishing House, 1998

Schumacher, E.F, *A Guide for the Perplexed*, Abacus, 1977

Scott, Mary, *Kundalini in the Physical World*, Routledge & Kegan Paul, 1983

Scrutton, Robert, *The Message of The Masters*, Jersey Neville Spearman, 1982

Seife, Charles, *Alpha and Omega, The Search for the Beginning and the End of the Universe*, Transworld, 2003

Shanks, Hershel and Witherington III, Ben, *The Brother of Jesus, The Dramatic Story and Meaning of the First Archaeological Link to Jesus and His Family*, Harper, 2003

Sheldrake, Rupert, *The Sense of Being Stared At and Other Aspects of the Extended Mind*, Hutchinson, 2003

Shermer, Michael, *Why People Believe Weird Things*, Henry Holt and Company, 1997

Sitchin, Zecharia, Genesis Revisited, *Is Modern Science Catching Up With Ancient Knowledge?*, Bear & Company, 1991

Sitchin, Zecharia, *When Time Began, Book V of the Earth Chronicles*, Avon Books, 1993

Stafford, Peter, *Psychedelics Encyclopedia, Third Expanded Edition*, Ronin, 1992

Stangroom, Jeremy, *What Scientists Think*, Routledge, 2005

Steiner, Rudolf, *Between Death and Rebirth*, Rudolf Steiner Press, 1975

Steiner, Rudolf, *Occult Science – An Outline*, Rudolf Steiner Press, 1963

Steiner, Rudolf, *The Effects of Spiritual Development*, Rudolf Steiner Press, 1978

Stemman, Roy, *Reincarnation, True Stories of Past Lives*, Piatkus, 1997

Stock, Gregory, *Redesigning Humans, Choosing our Children's Genes*, Profile, 2002

Stokes, Philip, *Philosophy – 100 Essential Thinkers*, Arcturus, 2003

Strong, James, *Strong's Exhaustive Concordance*, Baker Book House, 1983

Susskind, Leonard, *The Cosmic Landscape, String Theory and the Illusion of Intelligent Design*, Little, Brown and Company, 2005

Suzuki, Daisetz Teitaro, *An Introduction to Zen Buddhism*, Rider, 1969

Swedenborg, Emanuel, *Heaven and its Wonders and Hell*, The Swedenborg Society Inc, 1958

Sykes, Brian, *Adam's Curse, A Future Without Men*, Corgi, 2003

Sykes, Brian, *The Seven Daughters of Eve*, Corgi Books, 2002

Talbot, Michael, *The Holographic Universe*, Harper Collins, 1996

Tattersall, Ian, *Becoming Human, Evolution and Human Uniqueness*, Oxford University Press, 1998

Taylor, Timothy, *The Buried Soul, How Humans Invented Death*, Fourth Estate, 2002

Tellinger, Michael, *Slave Species of God*, Music Masters, 2005

Thompson, Mel, *Philosophy of Science*, Hodder & Stoughton Ltd, 2001

Thompson, Ernest, *Spiritualism in the Evolution of Philosophy and Spiritualism in the Evolution of Religion*, Two Worlds Publishing, 1950

Thompson, Thomas L, *The Messiah Myth*, Jonathan Cape, 2005

Von Daniken, Erich, *Chariots of the Gods?*, Souvenir Press, 1993

Von Daniken, Erich, *The Gods and Their Grand Design, The Eighth Wonder of the World*, Corgi Books, 1986

Von Daniken, Erich, *The Gods Were Astronauts, Evidence of the True Identities of the Old 'Gods'*, Vega, 2001

Von Daniken, Erich, *The Return of the Gods, Evidence of Extraterrestrial Visitations*, Element, 1998

Watson, Andrew, *The Quantum Quark*, Cambridge University Press, 2004

Watson, James D and Berry, Andrew, *DNA The Secret of Life*, Arrow Books, 2003

Watson, Lyall, *Supernature II, A New Natural History of the Supernatural*, Sceptre, 1986

Watson, Lyall, *Supernature, A Natural History of the Supernatural*, Sceptre, 1973

Watson, Lyall, *The Romeo Error, A Matter of Life and Death*, Coronet Books, 1974

Watson, Peter, *Ideas, A History From Fire to Freud*, Weidenfeld & Nicolson, 2005

Waugh, Alexander, *Time, From Micro-Seconds to Millennia – A Search for the Right Time*, Headline, 1999
Weinberg, Steven, *Dreams of a Final Theory*, Vintage Books, 1992
Wilber, Ken, *A Theory of Everything, An Integral Vision for Business, Politics, Science and Spirituality*, Gateway, 2000
Williams, Hywell, *Cassell's Chronology of World History, Dates, Events and Ideas that Made History*, Weidenfeld & Nicolson, 2005
Wilmut, Ian, Highfield, Roger, *After Dolly*, Little Brown, 2006
Wright, Robert, *The Moral Animal, Evolutionary Psychology and Everyday Life*, Abacus, 1996
Zimmer, Carl, *Evolution*, Arrow, 2003
Zimmer, Carl, *Soul Made Flesh, The Discovery of the Brain – and How it Changed the World*, Arrow Books, 2004
Zukav, Gary, *The Dancing Wu Li Masters*, Flamingo, 1984

Papers and Articles

Arkani-Hamed, Nima, Dimopoulos, Savas, Dvali, Georgi, *The Universe's Unseen Dimensions*, Scientific American, 2002

Begun, David R, *Planet of the Apes*, Scientific American, August 2005

Bekenstein, Jacob D, *Information in the Holographic Universe*, Scientific American, August 2003

Biello, David, *10th Planet Proves Bigger than Pluto*, Scientific American, 2006

Biello, David, *Supernovae Back Einstein's "Blunder"*, Scientific American, 2005

Biello, David, *Chimp Genome and First Fossils Unveiled*, Scientific American, 2005

Biello, David, *Colliding Clusters Shed Light on Dark Matter*, Scientific American, 2006

Biello, David, *Neuroscientists Probe Psychedelic Psilocybin*, Scientific American, 2006

Biello, David, *Scientists Predict Extinctions from Global Warming*, Scientific American, 2006

Biello, David, *Separation of Man and Ape Down to Gene Expression*, Scientific American, 2006

Biello, David, *Signs of Early Life in Australian Rocks*, Scientific American, 2006

Blanke, Olaf, Ortigue, Stphanie, Landis, Theodor, Seeck, Margitta, *Neurobiology: Stimulating own-body perceptions*, Nature, September 2002

Bousso, Raphael, Polchinski, Joseph, *The String Theory Landscape*, Scientific American, 2004

Caldwell, Robert R, Kamionkowski, Marc, *Echoes from the Big Bang*, Scientific American, 2002

Caldwell, Robert R, *Quintessence*, Physicsweb, Internet, 2000

Calvin, William H, *The Emergence of Intelligence*, Scientific American, 2006

Cann, Rebecca L, Wilson, Allan C, *The Recent African Genesis of Humans*, Scientific American, 2003

Close, Frank, *To Catch a Rising Star*, New Scientist, 2002

Crick, Francis, Kock Christof, *The Problem of Consciousness*, Scientific American, 2002

Damasio, Antonio R, *How the Brain Creates the Mind*, Scientific American, 2002

Dobbs, David, *Fact or Phrenology*, Scientific American, 2005

Duff, Michael J, *The Theory Formerly Known as Strings*, Scientific American, 2003

Fields, Douglas, R, *Beyond the Neuron Doctrine*, Scientific American, 2006

Ford, Lawrence H, Roman, Thomas A, *Negative Energy, Wormholes and Warp Drive*, Scientific American, 2003

Gazzaniga, Michael S, *The Split Brain Revisited*, Scientific American, 2002

Gould, Stephen J, *The Evolution of Life on Earth*, Scientific American, April 2006

Graham, Sarah, *Chandra Observations Confirm Existence of Dark Energy*, Internet, 2004

Graham, Sarah, *Scorched Remains Suggest Ancient Humans were Firestarters*, Scientific American, 2004

Greene, Brian, *The Future of String Theory*, Scientific American, 2005

Green, J. Timothy, *Near Death Experience, Shamanism, and the Scientific Method*, Internet, IANDS, 1998

Holt, Jim, *Beyond the Standard Model*, Scientific American, 2006

Ingham, Richard, *Chimp DNA Deciphered*, Cool Science, 2005

Ingham, Richard, *'Man of Flores' May Hold the Key to Evolution*, Sapa, 2004

Jogan, Craig J, Kirshner Robert P, Suntzeff Nicholas B, *Surveying Spacetime with Supernovae*, Scientific American, 2002

Kandel, Eric R, *The New Science of Mind*, Scientific American, 2006

Kane, Gordon, *The Dawn of Physics Beyond the Standard Model*, Scientific American, 2005

Kane, Gordon, *The Mysteries of Mass*, Scientific American, 2005

Kearns, Edward, Kajita, Takaaki, Totsuka, Yoji, *Detecting Massive Neutrinos*, Scientific American, 2003

Kokacy, A, *How Does Spending Prolonged Time in Microgravity Affect Astronauts*, Scientific American, 2003

Kole, William J, *Xena Fights for a Place in our Solar System*, Internet, 2006

Kostelecky, Alan, *The Search for Relativity Violations*, Scientific American, 2005

Kraft, Ulrich, *Unleashing Creativity*, Scientific American, 2005

Krauss, Lawrence M, *Cosmological Antigravity*, Scientific American, 2002

Krauss, Lawrence M, Starkman, Glenn D, *The Fate of Life in the Universe*, Scientific American, 2002

Lavine, Marc, *10th Planet Discovered*, Internet, 2005

Lavine, Marc, *US Astronomer Hails 'tenth rock from the sun'*, Internet, 2005

Leonard, William R, *Food for Thought*, Scientific American, 2003

Llewellyn Smith, Chris, *The Large Hadron Collider*, Scientific American, 2003

Logothetis, Nikos K, *Vision: A Window on Consciousness*, Scientific American, 2002

Long, Jeffrey, Bernstein, Paul, *Vision: A Window on Consciousness*, Scientific American, 2002

Lovell, Jeremy, *The Oldest Humans Just Became Older*, Internet, 2005

Luminet, Jean-Pierre, Starkman, Glenn D, Weeks Jefrey R, *Is Space Finite*, Scientific American, 2002

Maldacena, Juan, *The Illusion of Gravity*, Scientific American, 2005

McDonald, Arthur B, Klein, Joshua R, Wark, David L, *Solving the Solar Neutrino Problem*, Scientific American, 2005

Minsky, Marvin, *Will Robots Inherit the Earth?*, Scientific American, October 1994

Nash, Michael R, *The Truth and the Hype of Hypnosis*, Scientific American, 2001

Ostriker, Jeremiah P, Steinhardt, Paul J, *The Quintessential Universe*, Scientific American, 2001

Quinn, Helen R, Witherell, Michael S, *The Asymmetry between Matter*, Scientific American, 2003

Rees, Martin, *Exploring Our Universe and Others*, Scientific American, 2002

Rennie, John, *15 Answers to Creationist Nonsense*, Scientific American, April 2006

Samsel, W.T, *What is the Law of One?*, Internet, 2000

Sapa, *Amateur Stargazers Stumble Upon New Planet*, Internet, 2005

Schmid, Randolph E, *World's Oldest Jewelry Discovered in South Africa*, Cape Times, 2004

Shapiro, Ehud, Benenson Yaakov, *Bringing DNA Computers to Life*, Scientific American, 2006

Sinclair, David A, Guarente, Lenny, *Unlocking the Secrets of Longevity Genes*, Scientific American, 2006

Starkman, Glenn D, Schwarz, Dominik J, *Is the Universe Out of Tune*, Scientific American, 2005

Tattersall, Ian, *How We Came to Be*, Scientific American, 2006

Tattersall, Ian, *Once we Were Not Alone*, Scientific American, August 2005

Tattersall, Ian, *Out of Africa*, Scientific American, 2003

Tegmark, Max, *Parallel Universes*, Scientific American, May 2003

Thorne, Alan G, Wolpoff, Milford H, *The Multiregional Evolution of Humans*, Scientific American, 2003

Van Lommel, Pim, *About the Continuity of our Consciousness*, Internet, 2005

Van Lommel, Pim, *Medical Evidence for NDE's*, Skeptical Investigations, 2003

Van Lommel, Pim, Van Wees, R, Meyers, V, Elfferich, I, *Near-death experience in survivors of cardiac arrest: a prospective study in the Netherlands,* The Lancet, December 2001

Villarreal, Luis P, *Are Viruses Alive?*, Scientific American, 2004

Wayt Gibbs, W, *Ripples in Space Time*, Scientific American, 2002

Weinberg, Steven, *A Unified Physics by 2050?*, Scientific American, 2003

Wolfradt, Uwe, *Strangely Familiar,* Scientific American, 2006

Wong, Kate, *An Ancestor to Call our Own*, Scientific American, January 2003

Wong, Kate, *Chunk of Universes Missing Matter Found*, Scientific American, 2005

Wong, Kate, *Mini Human Species Unearthed*, Scientific American, 2004

Wong, Kate, *The Littlest Human*, Scientific American, February 2005

Wong, Kate, *The Morning of the Modern Mind*, Scientific American, June 2005

Wong, Kate, *Who Were the Neanderthals?*, Scientific American, 2003